T0183533

The Material Limits of Energy Transition: Thanatia

Alicia Valero · Antonio Valero · Guiomar Calvo

The Material Limits
of Energy Transition:
Thanatia

 Springer

Alicia Valero 🄳
Instituto CIRCE
University of Zaragoza
Zaragoza, Spain

Antonio Valero 🄳
Instituto CIRCE
University of Zaragoza
Zaragoza, Spain

Guiomar Calvo 🄳
Instituto CIRCE
University of Zaragoza
Zaragoza, Spain

ISBN 978-3-030-78532-1   ISBN 978-3-030-78533-8   (eBook)
https://doi.org/10.1007/978-3-030-78533-8

This Springer imprint is published by the registered company Springer Nature Switzerland AG
The registered company address is: Gewerbestrasse 11, 6330 Cham, Switzerland

Preface

Equilibrium thermodynamics is a revolutionary "young" science (it has only more than 170 years!) because it can change our understanding of planet Earth. It is a science that needs to be known by those who want to quantify the damage and degradation which humans are causing to the planet's capacity to support the human species. Engineers, physicists, chemists, geologists, environmentalists, ecologists, economists, forecasters and policymakers must learn from equilibrium thermodynamics and develop it further. This book focuses particularly on the abiotic (i.e. non-living) resources of our planet and on how they have been and will be affected by human behaviour.

Although this book presents a novel application of thermodynamics for assessing the Earth's mineral wealth, no expertise in the academic discipline of thermodynamics is required to understand the book's main message. A reader who is not well versed in thermodynamics can readily cherry-pick various parts of the individual chapters that (s)he finds illuminating.

Chapter 1 presents the context of the Earth's mineral resources and how this book proposes to assess their degradation.

Chapter 2 provides information on the extraction and use of energy and non-energy mineral resources from the past to the present. It focuses on key raw materials in the decarbonisation of the economy and describes some of the mineral criticality studies that currently exist.

While Chap. 2 focuses on economic demand, Chap. 3 addresses economic supply, analysing the availability of minerals on Earth. Besides, a description of the mining and refining processes of raw materials and the associated environmental and social impacts are presented.

After analysing mineral supply and demand, we outline in Chap. 4 the thermodynamic methodology proposed in this book. It describes a model of an economically degraded planet, Thanatia, used as a reference to assess the current state of mineral resources and their degradation velocity. In addition, the equations for calculating the exergy of mineral resources are provided, and thermodynamic rarity is proposed as an indicator of raw material criticality.

Using the thermodynamic tools presented in Chap. 4, Chap. 5 quantifies the exergy degradation of mineral resources on the planet since 1900. Various minerals' peak-production rates are assessed via Hubbert curves traditionally applied to fossil fuels. The novelty brought is that these can be represented in the same graph, taking into account the quantity and the quality of the resources. A study is also carried out on the mineral exergy balance of various regions of the world. With this approach, it is possible to detect the enormous inequalities between exporting and importing countries immediately. Finally, a monetary assessment of the exergy replacement costs of raw materials is undertaken. The adoption of such an accounting would imply a fairer appraisal of the mineral heritage of nations, taking into account future generations.

Chapter 6 focuses on assessing the potential raw material demand for the energy transition. Data are provided on the expected penetration of clean technologies, as well as on their material composition. Based on International Energy Agency or Greenpeace scenarios, the energy transition's exergy flows are analysed. It then becomes clear that there will be a shift from a dependence on fossil fuels to a multi-dependence on minerals, some of which very scarce, with extraction localised in only a few places of the world. Finally, some likely material bottlenecks in the development of clean technologies are identified.

Yet it is not only renewable energies that are mainly dependent on these scarce raw materials. So are new technologies that increasingly incorporate electrical and electronic components. Chapter 7 analyses these devices' thermodynamic rarity, focusing on probably the most resource-intensive technology: the vehicle.

Chapter 8 provides solutions to slow down the degradation of scarce mineral resources, showing how thermodynamics can help to manage the mineral wealth better. Thus, material substitution possibilities for various resource-intensive technologies are addressed. It also discusses the so-called circular economy and the thermodynamic limits it faces. Eco-design measures to increase the recoverability of raw materials at the end of life of products are also discussed. Finally, an insight into alternative mineral sources is presented: urban mining and asteroid mining.

We finally offer in Chap. 9 some reflections and conclusions drawn from our own research findings, claiming the need for a new humanism that cares about the future of the planet.

Here are some introductory remarks for readers versed in thermodynamics. Let us start with an example of the First Law of Thermodynamics: It is generally agreed that a calorie is a very small amount—just enough to raise the temperature of a gram of water by 1 °C. But if that gram of water were to carry a speed of 329 km/h, it would have the energy, now kinetic, of a calorie! This is surprising because we have not internalised the concept of energy. Moreover, we seem to associate energy with damage rather than heat. Yet in reality, a punch from a boxer can communicate less energy than a gentle caress. Numbers say nothing if they are not internalised. As Protagoras said, "man is the measure of all things", but be careful; the sense of physical damage is not an appropriate measure of deterioration.

Despite these paradoxes, we can make statistics and foresight studies also because energy is additive. We can add electrical energy to thermal energy and to any other energy manifestations without making mistakes, as long as we distinguish between primary and final energy.

However, what is no longer straightforward is to understand the second law of thermodynamics quantitatively. If energy is not lost, where does it go? We know that heat cools, metals rust, the wind stops, water rains, living beings age and die, and the planet degrades. But how fast does Earth degrade? And how fast does it regenerate? If the planet is finite, how long will it take until its exhaustion? And how can we stop this degradation? These are the questions that currently have no scientific answers. There are proclamations, considerations, predictions, but there is no quantitative science behind it. We need a transdisciplinary theory based on thermodynamic criteria, which goes beyond it. We need a science that builds ever more detailed statistics, even if these are initially based on imprecise and fuzzy but objective data, that serves as a rudder and thermometer of the damage inflicted by our civilisation on planet Earth.

From the second law, we know that sooner or later, all quality energy will become heat. Heat is the sink of all energies, so the energy we receive every day from the Sun moves the biosphere. Yet, unfortunately, humankind degrades natural resources faster than the Sun replenishes them. If any degradation can be measured with entropy, we need to focus on understanding what entropy is.

We can easily understand that if energy is conserved and hot bodies cool spontaneously, isolated systems tend to increase their entropy. Unfortunately, entropy has units of energy divided by temperature, making it complex to comprehend and impractical to use. First of all, entropy is not a property that behaves linearly like energy. Losing 1 °C at 5505 °C (i.e., at 5778 Kelvin, the equivalent temperature of solar radiation) is not the same as losing it at 27 °C (300 K) or losing it at −73 °C (200 K). In other words, entropy forces us to live with exponential behaviour, which is difficult to understand for those not used to mathematical thinking. On the other hand, using units such as kWh/K does not facilitate the quantitative explanation of the social consequences of degradation. Therefore, it is not surprising that entropy is often used as a metaphor, moving away from quantitative messages.

The solution to these issues comes with exergy. Exergy is more interesting than entropy because it simultaneously integrates the First Law of Thermodynamics, energy conservation, and the second, the entropy law. In other words, exergy simultaneously condenses information about energy and entropy. Mathematically, it has a straightforward formula: the change in energy minus the ambient temperature multiplied by the entropy change. Its generic formula is:

$$B = \Delta E - T_0 \Delta S$$

where B is the exergy, ΔE is the energy change with respect to the reference, T_0 is the absolute temperature of the reference and ΔS is the entropy change with respect to the reference. It is therefore easy to see that the exergy property integrates both energy and entropy. It is measured in energy units and is additive, which makes it much more practical and easy to understand. Technically, exergy measures the maximum work obtained from a system when it is brought into equilibrium with the environment. Alternatively, exergy represents the minimum work necessary to bring the system from equilibrium with the environment to a given alternative state.

Note that to define exergy, we have added a new concept, the reference environment, which can open up a new problem rather than providing a solution depending on how we see it. The reference environment is not originated from the convenience of calculations but from observing the physical behaviour of matter. It is the ground if we speak of a ball falling down, it is the absence of wind if we speak of the atmosphere, it is the diluted CO_2 in the environment if we speak of a fossil fuel that has been burned, it is rusted metal, it is a dilution of pollutants in the sea and the atmosphere, it is the unavoidable dispersion of materials throughout the crust, it is the irretrievable loss of natural resources, and it is death. It is Thanatia, a planet easily imaginable if we observe Nature's degradation, at temperature T_0, slowly increasing if we do not stop climate change.

Thanatia's message flips the way the degradation of natural resources is perceived and assessed. Instead of moving from today to a defined temporal future, Thanatia's thinking suggests time to run backwards. If we accept an end, i.e., the finitude of resources, we can ask ourselves how fast we are approaching it. It is as if we had to take a flight at a fixed date and time. We organise our time backwards, we prepare the luggage, commutes, and all the necessary steps to arrive on time to take the plane. In short, it is forward vs. backward thinking. This change in thinking helps us to find a way to avoid any pessimistic future.

This is, dear reader, what this book is about. It shows that equilibrium thermodynamics can explain how relentless loss of the planet's mineral wealth—a loss which the energy transition will accelerate—can be assessed. However, now we are no longer talking about the equilibrium between bodies as classical thermodynamics does, but about the equilibrium between humans and the planet, which is why the word equilibrium thermodynamics takes on a new nuance. Perhaps to avoid confusion it should be called the thermodynamics of sustainability.

Our work on this topic started in 1998 with several papers and a book entitled "Desarrollo Económico y Deterioro Ecológico" (meaning "Economic Development and Ecological Deterioration"). After three Ph.D.s and more research papers, our studies led us in 2014 to write a book entitled, *Thanatia. The Destiny of the Earth's Mineral Resources. A Thermodynamic Cradle to Cradle Assessment*. Now, seven years later, after five additional Ph.D.s and more than 50 scientific papers, we present this new book, opening up new questions on a crucial issue for twenty-first-century humankind: the conservation and rational management of the planet's mineral resources for future generations.

The authors thank the Spanish Ministry of Economy, Industry and Competitiveness for the funding received to write this book through Project ENE2017-85224-R.

Zaragoza, Spain   Alicia Valero
                                                                                          Antonio Valero
                                                                                          Guiomar Calvo

Contents

About the Authors

Alicia Valero studied chemical engineering at the University of Zaragoza (Spain), where she also completed a master's degree in energy efficiency and industrial ecology. In 2008, she obtained a European PhD from the University of Zaragoza. She is currently an associate professor in the Department of Mechanical Engineering (University of Zaragoza) and head of the industrial ecology group at the CIRCE Institute. Her research activity has focused on the exergy evaluation of the Earth's mineral capital, a subject in which she has been working for 15 years. She has received four international awards. She is the co-author of over 50 publications in scientific journals and book chapters and more than 60 communications to international congresses. She has participated in more than 30 national and international projects related to the study and optimisation of energy and materials. She belongs to various international experts' committees on raw materials.

Antonio Valero is the chair in thermal systems at the University of Zaragoza (Spain). He is the director and founder of the Research Centre for Energy Resources and Consumption (CIRCE Institute) belonging to the University of Zaragoza (Spain). Since 1986, when he published the general theory of exergy saving, he has developed various thermodynamic theories, including thermoeconomics and exergoecology, used for the optimisation and evaluation of natural resources. He has directed more than 35 Ph.D.s and has co-authored hundreds of scientific papers, book chapters and communications to conferences on these topics. He is a fellow member of the American Society of Mechanical Engineers. He received the ASME James H. Potter Gold Medal Award 1996 for advancing the theory of thermoeconomics to a new level, as well as the Stanislaw Ocheduszko Medal 2016 to distinguish his contributions to thermodynamics, among other international recognitions.

Guiomar Calvo graduated in geology from the University of Zaragoza (Spain) in 2010. She studied a master in introduction to research in geology (2011) and a master in eco-efficiency and industrial ecology (2013) at the same university. In 2016, she defended her doctoral thesis, entitled "Exergy assessment of mineral extraction, trade, and depletion", which consisted of the evaluation of mineral resources and mineral depletion from a thermodynamic point of view. She is the co-author of over 50 scientific papers, conference communications and book chapters, along with three dissemination books related to minerals. She has participated in various national and European research projects related to assessing and optimising raw material use. She has worked as a postdoctoral researcher at CIRCE Institute, where she has carried out the vast majority of her research activity. She has also worked as a lecturer at the International University of La Rioja (Spain).

List of Figures

List of Tables

Chapter 1
What Is This Book About?

Abstract Humankind has relied on the extraction of different raw materials for centuries, starting with iron, copper or gold to a large number of metals and fossil fuels currently used in multiple sectors, thanks to technological development. Still, this change has also led to other issues, such as increasing CO_2 at a global level and climate change. One way to mitigate these problems is to rely on renewable energy sources that use the Sun or wind to generate electricity instead of burning fossil fuels. However, these technologies need certain elements that are scarce on the planet or very complicated to extract. To assess our planet's mineral loss, in this book, we will use thermodynamics, specifically its second law, that will allow us to explain this degradation process physically. Using Thanatia as a baseline, a hypothetical land where all concentrated materials have been extracted and dispersed, and all the fossil fuels have been consumed, we can assess the cost of replacing minerals through a grave-to-cradle approach and combine it with the more traditional cradle-to-grave approach.

Everything around us is made up of minerals. Dozens of chemical elements are used in smartphones, household appliances, vehicles, concrete, paints, detergents, etc., that come from the extraction and processing of these minerals. We start from the advantage that the natural processes that have been taking place over millions of years on our planet have been concentrating these elements in the form of mineral deposits. Mining becomes then our primary source, from where we extract the minerals that we then use. Since these mines are not infinite, it is legitimate to ask what limitations may exist in the short, medium and long term.

The increase in population, globalisation and the change in consumption trends are causing the use of resources to increase dramatically every year. In fact, the primary extraction of quarry products, metallic minerals, fossil fuels and biomass increase year on year. On a limited planet, are we going to be able to maintain this pace forever? What consequences will this have on future generations and on the planet?

Historically, the extraction and use of raw materials have been closely linked to human development. We have gone from consuming about 3 kg of natural resources per inhabitant per day in prehistory to 44 kg in our current industrialised society

© The Author(s), under exclusive license to Springer Nature Switzerland AG 2021
A. Valero et al., *The Material Limits of Energy Transition: Thanatia*,
https://doi.org/10.1007/978-3-030-78533-8_1

(Friends of the Earth, 2009). Our prehistoric ancestors obtained mineral resources through surface collection, selecting those materials most suitable to serve as cutting tools, such as quartzite or flint. Other readily available materials have historically been used as cosmetics and for decorative purposes. The Egyptians used mixtures of oils with dust from the crushing of lead minerals, such as galena, and copper, such as malachite, among others, to make *kohl*, a thick black substance that they later applied to outline their eyes (Hallmann, 2009).

With the emergence of more complex societies, mining became much more relevant, using materials for own consumption and exchange. Different metals gradually gained more weight, including copper, bronze (an alloy of copper and tin), and gold, highly desired both for ornamentation and jewellery and for its economic value.

A well-known example globally is the ancient gold mine of *Las Médulas*, located in the province of León (Spain), considered the largest open-pit metal mine in the Roman Empire (Fig. 1.1). The exploitation was carried out by the force of water, with the method known as *ruina montium*. Water was channelled and accumulated at the top of the mountain and, as this water was released through steep galleries, and by the force of gravity, the mountain would erode, dragging the gold to the washing sites located at the bottom (Pérez García et al., 1998). It is estimated that the Romans were able to extract between five and seven tons of gold from this location, which has left as an inheritance the characteristic landscape that this area presents. Such is the value of this natural space that UNESCO included it as a World Heritage Site in 1997.

Fig. 1.1 Ancient gold open-cast exploitation of the Roman Empire of *Las Médulas* (Castilla y León, Spain). *Author* Rafael Ibáñez Fernández. GNU FDL. Wikimedia Commons

Historically, gold that appears in its native state has also been mined manually using pans. This technique, widespread in past centuries, consisted of using a pan filled with sand and immersed in water; through a series of circular movements, and due to the difference in density of the materials, the gold deposited at the bottom while the gravel was washed off (Fig. 1.2). This same technique was also used during the gold rush in the United States in the middle of the nineteenth century, along with the sluice boxes, where the material was washed. During this time, dry gold washing also became popular, driven by the lack of water in many regions. In this case, the mineral was deposited inside a conical wooden pan. Throwing the material into the air, lighter materials dispersed leaving the heavier ones at the container's bottom. However, as can be assumed, this was not a very effective method since only large gold nuggets could be recovered (Taylor Hansen, 2007). The use of pans and decantation in artisanal gold mining continues to this day.

The technological development that has taken place over the centuries has progressively increased the number of metals and other elements that are used, from just a few in the seventeenth century to practically all of those contained in the periodic

Fig. 1.2 Engraving from the work of Georgius Agricola, *De re Metallica*, published in 1577, representing gold extraction techniques in Germany in the sixteenth century. The sluice boxes ensured that gold, a denser material, accumulated in the channels. There is also a person panning, a traditional method still used in some places

table today. This is even more evident in the case of elements used in the energy sector (Zepf et al., 2014). Initially, the materials necessary to manufacture mills that harnessed the energy of the wind were few: chiefly iron, wood and stone; the same occurred with candles or oil lamps used for lighting. With the industrial revolution and the steam engine's invention, other elements were introduced in the energy sector: copper, tin, lead, manganese, etc., but they were still few in number. The appearance of motor vehicles changed the situation drastically again, increasing not only the consumption of fossil fuels but also that of other metals that until now had not been very useful.

Today, we use many elements in different applications that increase our convenience and comfort. For instance, in a smartphone, we can find several dozen elements of the periodic table, which include tin and indium oxide in the touchscreen and rare earth elements that produce the colours we see and, of course, lithium in batteries (Merchant, 2017).

Electricity generation is no exception either, since it requires large amounts of elements, some of them very valuable and scarce, to produce wind turbines, photovoltaic panels, etc. For example, to produce one gigawatt (GW) of electrical power equivalent to that which a natural gas-fired power plant could supply would require a total of 200 5-megawatt (MW) wind turbines or 1,000 1-megawatt (MW) wind turbines. This would imply the use of approximately 160,000 tons of steel, 2,000 of copper, 780 of aluminium, 110 of nickel, 85 of neodymium and 7 of dysprosium for its construction. These are not negligible amounts if it is estimated that in the future the energy produced by wind turbines in 2050 could be around 2,200 GW (International Energy Agency, 2019).

Worse still, as can be seen in Fig. 1.3, wind turbines are one of the renewable technologies that require the least variety of elements for their production, but others such as the electric car employ over 40 different elements, and that's before considering the rest of the necessary materials such as plastics, glass, polymers, etc. (Valero, 2018).

Considering the intense use of materials from clean technologies, will the deployment of renewable energy required to achieve the Paris Agreement goal (preventing Earth's temperature rise of over 2 °C before the end of the century) be possible? We want to move from a society based on non-renewable energy sources to one based on renewable sources. However, what has been rarely considered is that these technologies require a greater diversity of materials than conventional energy sources and that, in addition, they are highly voracious in many different elements.

As we currently know, society is completely dependent on many elements, almost all of which come from the primary extraction of certain minerals. In our society, no product exists that does not contain minerals or whose production does not directly involve minerals. Consequently, the global extraction of natural resources has increased exponentially, as can be seen in Fig. 1.4, and the same situation can be observed for other materials.

The amount of biomass that has been extracted, comparing 1900 and 2017 data, has increased fivefold, in the case of fossil fuels 15-fold, and by a factor of 43 and 65 in the case of metallic and construction minerals, respectively (International Resource

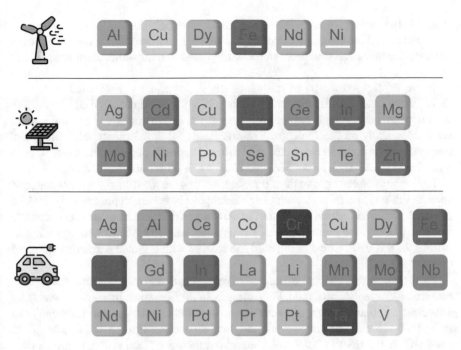

Fig. 1.3 Some of the elements that are used to manufacture clean technologies (Valero et al., 2018)

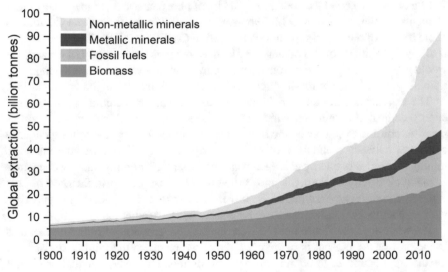

Fig. 1.4 Global material extraction from 1900 to 2017 in billions of tons (International Resource Panel, 2019)

Panel, 2019). In fact, so far in the twenty-first century (in the last 20 years) we have extracted almost the same amount of copper that was extracted in the entire twentieth century, and this same situation can be extrapolated to many other elements (USGS, 2018).

However, this extraction of raw materials is not equally distributed across the globe. In the case of mineral resources, it is geology that conditions the places where the elements have been concentrating over time. In Australia, for example, there are economically profitable deposits of practically all the elements, while in Spain, despite having a considerable amount of mineral deposits of different elements, only a few basic metals such as copper, lead or zinc can be economically extracted.

If we take as an example some of the elements that are most crucial to our economy, such as lithium, which is essential for electric car batteries, approximately 55% of the total global extraction originated in Australia in 2019. Another representative example of that same year are rare earth elements, used in many technological applications; in this case, China dominated the market with a global extraction quota of over 60% (USGS, 2020).

Furthermore, this unequal extraction of resources is associated with consumption that is also unevenly distributed. For example, in Europe, three times more resources are consumed than in Asia, and four times more than in Africa, and someone born in the United States consumes even more than an average European. For example, a child born in the USA in 2019 will, throughout their life (78.6 years), require a total of 9,129 kg of iron, 937 kg of primary aluminium, 444 kg of copper, 432 kg of lead, 211 kg of zinc, 13,693 kg of salt and 6,503 kg of phosphate rock, among many other elements, in addition to some 1,800 barrels of oil, 150 tons of coal and 7.7 million cubic meters of natural gas (Minerals Education Coalition, 2019). This implies that if all the inhabitants of the planet tried to live today as an average US citizen, we would need to multiply the current copper extraction by two to cover the demand of a single year and something similar would happen with the rest of the raw materials.

The exponential extraction of materials also entails an increase in the required energy dedicated to mining, which in turn can significantly impact the environment. According to studies by the International Energy Agency, the mining industry consumes between 8 and 10% of global energy. As an example of how intensive mining is in terms of energy use, each year, the Australian mining industry consumes as much electricity as Portugal, and if the cost of transport is also factored in, it is equal to the energy consumed in Spain. It is clear that there can be no materials without energy, but neither can there be energy without materials.

So, what does the future hold? Knowing the consumption of mineral resources in the past or the present is relatively simple: we resort to the mining statistics of the different countries to obtain approximate figures. However, of equal or greater importance is trying to predict what future behaviour will be to anticipate eventual shortage problems. To this end, different models have been created based on statistical calculations and trend analysis, among others. Some striking insights can be gleaned from these studies. In the case of silver, gold, copper or nickel, their demand is estimated to increase fivefold by 2050. Taken alone, this figure doesn't provide much value but compared to the known amount of these elements in mines today, it exceeds by far the amount that would be possible and profitable to extract at current prices (Halada et al., 2008). This implies that we will end up needing much more than the Earth can provide and that the true costs of the depletion of non-renewable natural resources, and the consequent degradation of ecosystems, will be much greater and will steadily increase if this trend continues.

In short, the planet Earth has become an enormous mine. Not only are many elements extracted, but millions of tons of sterile rock are generated that accumulate around the mines. In 2000, mineral processing and refining were responsible for around 12% of sulfur dioxide emissions, a gas that is the main cause of acid rain (Smith et al., 2004). To this figure, we must add the landscaping impact of the mines, the impact on land use and the possibility of accidents such as the one that occurred in Brazil in early 2019 where an iron ore containment dam collapsed causing serious damage, both in loss of human lives as in material and environmental terms.

Indeed, nature is no longer as abundant as before; if it were, perhaps the contribution of mining would not be as devastating. However, the intense technological development of the twenty-first century is forcing society to react, each of us becoming increasingly aware of the detrimental effects our actions are having on the planet.

Another reaction to exponential extraction is increased global awareness of the fact that we are experiencing a loss of natural capital that will never be regenerated. In other words, there are limits to growth. In fact, many decades ago this was evidenced by the report to the Club of Rome of the book *The Limits to Growth* (Meadows, 1972), based on a report carried out at the Massachusetts Institute of Technology, a book that was updated in 2004 (Meadows et al., 2004). It predicted that population growth, coupled with the exponential use of fossil fuels and minerals, would bring the world to the brink of collapse in just a few generations. This prediction became increasingly real, especially with the subsequent oil crisis in the 1970s and 1980s, only to be forgotten during the bubble boom of the late 1990s and early 2000s. At that time, global growth was driven by the idea that the planet could absorb all the environmental impacts caused by social development, that mineral and energy resources were sufficient to maintain unlimited growth, and that innovation and technological development would be able to solve existing or future problems.

Looking back, the lack of accurate data generated by those early predictions of future scarcity does not weaken the Club of Rome message. Environmental problems worldwide have been worsening and the environmental footprint generated by the planet's population is already exceeding the limits of what the Earth can support.

Earth Overshoot Day is the day in which all the resources that the Earth can regenerate in a year have been consumed (Fig. 1.5). That is, if we maintained a sustainable rate of consumption and generation of waste that our planet was capable of absorbing, this day would occur on 31 December. Already in 1970, this overcapacity limit was reached on 23 December, not too far from the ideal date; however, in 2019 the limit of the planet was reached on 1 August. In 2020, this day was pushed back three weeks, until 22 August, mainly due to the global COVID-19 pandemic, which slowed down the consumption of materials, reaching levels similar to 15 years ago (Earth Overshoot Day, 2020). What does this mean? Essentially it means that each year we need around 1.75 planets to maintain our growth rate, but as we all

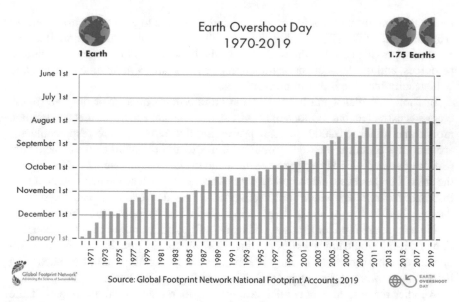

Fig. 1.5 Evolution of Earth Overshoot Day from 1970 to 2019 (Global Footprint Network: www. footprintnetwork.org)

know, we only have one (WWF, 2018). In addition, due to the unequal distribution of consumption, if all the inhabitants of the planet wanted to live at the level of an average Spaniard, it would take two-and-a-half planets, a figure that would increase to five planets if we all wanted to live at the same level as an American.

The environmental footprint of each one of us continues to increase, and it does so even more considerably in developing countries, which seek to raise their quality of life to the level of developed countries. The expected large deployment of renewable energy will also affect the type of materials that are extracted, producing a shift from dependence on fossil fuels to dependence on scarce materials, as we will see later. Although many continue to deny the existence of materials shortages and the need to address them from an economic point of view, the fact is that physical exhaustion is related to the limited amount of natural resources that the Earth has to offer. If this threat is to be taken seriously, we must begin to rigorously account for what raw materials are produced, from their primary extraction to the manufacture of the final product, and then from their disposal to full dispersal.

Such monitoring could be carried out through life cycle assessment (LCA), a widely established methodology to evaluate the environmental footprint and the associated environmental impacts from the start of a good or service, the so-called cradle, until the end of its life, the grave. This cradle-to-grave analysis, and in the opposite direction, can be extended to cover the entire planet, and when applied to the extraction of mineral resources the loss of mineral wealth caused by human actions can be analysed.

With this idea in mind, thermodynamics, specifically its second law, allows us to explain any degradation process in a physical way. The ultimate goal is to be able to carry out a thermodynamic analysis of the depletion of raw materials on Earth through the entire life cycle, that is, from cradle-to-grave, and then from grave-to-cradle. This cyclical process occurs naturally in ecosystems if sufficient time elapses, thanks to the action of the Sun. However, the Sun has little influence on the cycle of minerals, and once they enter the technosphere (the system formed for all the elements created by human beings), they end up dispersing, rarely being available for reuse.

The mineral cycle is much more complex, and humans are shortening it at will. The minerals, concentrated in deposits, can be considered nature's great repository. This warehouse, once it is subjected to extraction and processing, becomes part of the industry. After the products are manufactured, the minerals already belong to the inventories (or stock) that are in use, and when they reach the end of the useful life of these products, that is, when society no longer needs them, they usually end up in landfills or, at best, recycled. In landfills, the elements end up dispersing, making their recovery practically impossible and, even in the case of recycling, full recovery cannot be achieved. Therefore, instead of a cycle as such, the life of these materials can be represented through a spiral that never closes, since even applying the best recycling techniques, it is not possible to recover all the material used for its manufacturing due to thermodynamic limitations.

Today, large amounts of minerals are mined without considering this dispersion process, something that could be avoided if studies on recovery costs were carried out both now and in the future. Knowing, or at least having an order of magnitude of, what the costs of trying to close these cycles are, that is, estimating the cost of reconcentrating those minerals dispersed by humans in a deposit, can be crucial. It could help to clarify the size of the problem. Only in this way will we understand the need to adequately manage the scarce resources that nature provides us without asking for anything in return.

In a planetary cradle-to-grave study, and then from grave-to-cradle, it is necessary to clearly identify what we mean by cradle and what we mean by the grave. The cradle is simple to determine in this case: it refers to the repository where the minerals are initially, which have been concentrated over time through natural processes, that is to say, the mineral deposits and the mines. Extrapolating this idea to all the mineral resources that humans use, the cradle is then the Earth as we know it today. The grave, however, is somewhat more complex to define. When analysing the life cycle of a material, the grave is normally the dump where they end up; by establishing a simile, on a planetary level, the grave could be the final dump where all resources have been irreversibly dispersed, that is, a planet commercially dead in resources. In order to develop this hypothesis of a planetary grave, it is necessary to create a model capable of adequately representing the commercial purpose of the planet, hence the *Thanatia* model (Valero & Valero, 2014).

The word *Thanatia* comes from the Greek *Thanatos*, meaning death, and represents a hypothetical land where all concentrated materials have been extracted and dispersed throughout the crust and in addition, where also all the fossil fuels have been consumed (Fig. 1.6).

This crepuscular planet would have an atmosphere, hydrosphere and continental crust with a determined composition different from the current one. For example, since there are no concentrated mineral deposits in it and all the fossil fuels have been burned, the atmosphere would have higher CO_2 concentrations than current ones. Similarly, almost all the water available in the hydrosphere would have a different composition as the fresh water (representing only 3% of the planet's total water) would be mixed with salt water. This model does not mean that *Thanatia* is the end of life for our planet; it only implies that resources would no longer be available in their current form, concentrated and ready for extraction.

Fig. 1.6 Conceptual hourglass of the passage from *Gaia* (our planet today) to *Thanatia* (planet with dispersed and/or totally consumed natural resources)

Starting from *Thanatia*, it is possible to assess the cost of replacing minerals through a grave-to-cradle approach, since this cost would be that corresponding to the useful energy needed to re-concentrate those minerals to the concentration that appears in their respective deposits, that is, moving from a dispersed environment like *Thanatia*'s to the concentration of current mines. To have a unified vision and make comparisons between different resources and countries, exergy is used as a unit of measurement, which also helps us take into account not only the quantity but also the quality of these resources.

It is essential to keep in mind that the quality of mineral resources is not always the same, that is, it makes no sense to add a ton of iron to a ton of gold. It would be like comparing apples to pears since the processes involved in extraction and processing are not comparable at all. Because of this exergy, and specifically, exergy replacement costs, consider the terms not only of quantity but also of quality, which allows us to compare apples to apples. Accordingly, the mineral loss caused by the past extraction and the one to come can be evaluated using exergy and the *Thanatia* model.

All the concepts involved in this process, the materials cycle, the life cycle assessment of mineral resources, exergy, replacement costs, etc., will be explained in detail throughout the book using applications and practical cases, focused on the thermodynamic evaluation of mineral resources, the energy transition and technologies.

Given the current trends in the extraction of materials, the depletion of mineral resources does not seem to be a priority, but to prevent our planet from becoming *Thanatia*, we need to act now.

References

Earth Overshoot Day. (2020) *Earth overshoot day*. https://www.overshootday.org.

Halada, K., Shimada, M., & Ijima, K. (2008). Forecasting of the consumption of metals up to 2050. *Materials Transaction, 49*(3), 402–410.

Hallmann, A. (2009) Was ancient Egyptian kohl a poison? In: J. Popielska-Grzybowska, O. Białostocka, & J. Iwasczuk (Eds.) *Proceedings of the third central European conference of young Egyptologists* (pp. 69–72). Warsaw.

International Energy Agency. (2019) *World energy outlook 2019*. https://www.iea.org.

International Resource Panel. (2019) *Global material flows database*. https://www.resourcepanel.org/global-material-flows-database.

Meadows, D. H., Meadows, D.L., Randers & J. Behrens, W.W. (1972). *The limits to growth*. Universe Books.

Meadows, D. H., Randers, J., & Meadows, D. L. (2004). *Limits to growth: The 30-year update*. Chelsea Green Publishing.

Merchant, B. (2017). *The one device: The secret history of the iPhone*. Little Brown, 416 pp.

Minerals Education Coalition. (2019). *Minerals education coalition*. https://mineralseducationcoalition.org.

Pérez García, L.C., Sánchez-Palencia, F.J., Fernández Manzanos, J., Orejas Saco del Valle, A., & Fernández Posse, M.D. (1998). Las Médulas (León), la formación de un paisaje cultural minero. *Boletín Geológico y Minero, 109*, 5–6, 157–168.

Smith, S. J., et al. (2004). *Historical sulfur dioxide emissions 1850–2000: Methods and results.* Available at: https://www.pnnl.gov/main/publications/external/technical_reports/PNNL-14537. pdf.

Taylor Hansen, L. D. (2007). La 'fiebre del oro' en Baja California durante la década de 1850: Su impacto sobre el desarrollo del territorio. *Región y Sociedad, 19*(38), 105–127.

USGS. (2018) *Mineral commodity summaries 2018. United States geological survey.* Available at: https://minerals.usgs.gov/minerals/pubs/mcs/2018/mcs2018.pdf.

USGS. (2020). *Mineral commodity summaries 2020. United States geological service.* Available at: https://pubs.usgs.gov/periodicals/mcs2020/mcs2020.pdf.

Valero, A., Valero, A., Calvo, G., Ortego, A., Acaso, S. & Palacios, J.L. (2018). Global material requirements for the energy transition. An exergy flow analysis of decarbonisation pathways. *Energy, 159*, 1175–1184.

Valero, A., & Valero, A. (2014). *Thanatia: The destiny of the Earth's mineral resources: A thermodynamic cradle-to-cradle assessment.* World Scientific Publishing Company.

WWF. (2018). *Living planet 2018: Aiming higher.* Available at: https://www.wwf.eu/campaigns/ living_planet_report_2018/.

Zepf, V., Reller, A., Rennie, C., Ashfield, M., & Simmons, J., BP (2014): Materials critical to the energy industry. An introduction. 2nd edition.

Chapter 2
The Mineral Voracity of Human Beings

Abstract Fossil fuel and mineral demand have considerably increased in the last few decades, even reaching an exponential trend. Oil and natural gas consumption accounts for more than half of the total demand in recent years, the Middle East and Saudi Arabia being the main producing regions. Regarding unconventional fossil fuels, the United States is the country that produces the most, with almost 40% of its gas production coming from this type of deposits. As for minerals, iron, aluminium, gypsum and limestone have experienced great growth, as well as phosphate rock, mainly used in the food sector. Some of these non-fuel minerals, such as rare earth elements, cobalt, lithium, niobium, tantalum or gallium, which are considered critical by different regions, have seen how the appearance of new applications in technological sectors has increased interest in their exploration and extraction. In this chapter, we will explore the extraction trends of each element and its uses in our society. Last, we will talk about different criticality studies and which substances are considered as such.

Raw materials are essential not only to produce a wide variety of products and services used on a daily basis but also for the development of renewable energy and new technologies, in addition to a number of as yet unknown future applications. Our economic system is closely linked to the extraction of raw materials, including fossil fuels and minerals, which we will look at in detail below. However, it is important to specify that, in many cases, statistics speak of production instead of extraction as such. In this book, for the sake of unifying criteria, we have, generally speaking, decided to talk of extraction, referring to what is in the Earth's crust and is taken out of it for the benefit of human beings. It is important to analyse, initially, what type of materials we demand before calculating how much is available to establish the basis of the issue of natural resources depletion.

How has historical demand for the various energy and non-energy minerals evolved? What is the origin and destination of these resources? What are they used for? Which sectors are the most voracious in minerals? What are the most critical and strategic raw materials for the development of economies? This chapter will attempt to address these issues.

2.1   Demand for Fossil Fuels

Consumption of coal, oil and natural gas helps maintain the current state of civilisation. We use them for heating and cooling, also for transport, both private and for goods. They are also one of the main sources of electricity generation, apart from the entire petrochemical industry that manufactures plastics, textiles, lubricants, asphalts and others.

Figure 2.1 shows the historical evolution of the extraction of fossil fuels worldwide. Although the use of coal as an energy source predominated throughout much of the twentieth century, it has gradually been giving way to oil and natural gas, leaving the extraction of the three types of fossil fuels today rather equalised.

Fossil fuels are by far the most significant energy source used today, accounting for 85% of global primary energy consumption, while the remaining 15% is distributed between hydroelectric, nuclear and renewable energy (Fig. 2.2).

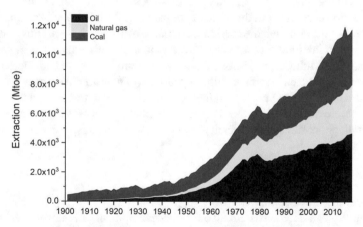

Fig. 2.1 Extraction of fossil fuels from 1900 to 2018. Data in millions of tons of oil equivalent (Mtoe). *Data sources* USGS, BP

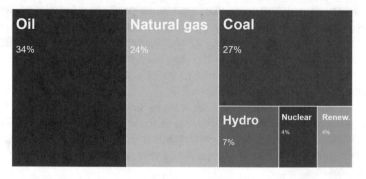

Fig. 2.2 Primary energy consumption in 2018 by type of energy source. *Data source* BP (2019)

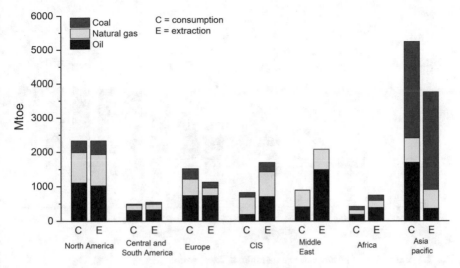

Fig. 2.3 Consumption and extraction of fossil fuels by country in 2018. *Data source* (BP, 2019)

By region, the area that consumed the most fossil fuels during 2018 was the Asia Pacific, with almost 5,300 Mtoe, which includes countries such as Australia, China, India and Japan; it was also the region that extracted the highest amount compared to the others (Fig. 2.3). Among the regions that consume more fossil fuels than they extract, we also find Europe, which consumes almost three times more natural gas than it extracts.

Among the regions that extract the most, the Middle East stands out, with Saudi Arabia at the top, being an area that does not extract much coal, but does extract large amounts of oil, specifically almost 1,500 Mtoe (BP, 2019). To a lesser extent, the same is true of the Commonwealth of Independent States (CIS), Africa, and Central and South America.

To the extraction of conventional fossil fuels, we must add that of unconventional ones, those extracted by much more aggressive techniques, a well-known example of these techniques being *fracking*. This technique is fundamentally based on injecting a certain volume of liquid into the subsoil at high pressure, which generates fractures and cracks in the rocks, causing the gas or oil that may exist in them to be released.

The extraction of natural gas from unconventional sources in recent years is shown in Fig. 2.4, with shale gas being the most productive, followed by tight gas and CBM (coalbed methane). Although the availability of methods to recover this gas has led to greater access to these resources, they can entail a series of environmental and social problems. These include possible effects such as water pollution or increased earthquakes in areas where this technique is used, as well as the impact on the natural landscape, among others.

The United States is the country that produces the most unconventional fossil fuels, with almost 40% of its gas production coming from this type of deposits. The trend of gas extraction of this type has been increasing over the years. However, at

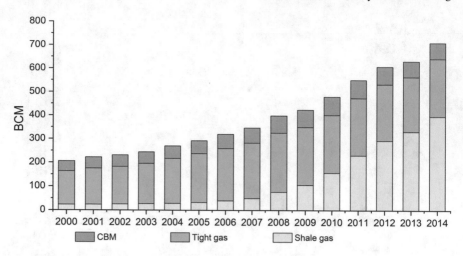

Fig. 2.4 Extraction of unconventional natural gas. *Data source* IEA

the end of 2015, the extraction of oil through this system did begin to decrease and by the end of 2016, it had decreased by more than 15% (U.S. Energy Information Ad, 2019). That said, it resumed in 2017 with the arrival of Donald Trump at the White House and this uptick continues to this day.

This trend observed in the case of fossil fuels, with an almost exponential growth in extraction, also occurs in the case of non-energy mineral resources (hereinafter, mineral resources), as we will see below.

2.2 Demand for Minerals

Mineral resources can be broadly classified into two groups: metallic minerals and industrial minerals. Metallic minerals include all those minerals from which metals are recovered, such as iron, copper, silver, gold, lead, zinc, etc. On the contrary, industrial minerals are those that are mainly used in industrial processes, directly or indirectly, to take advantage of their physical and chemical properties, and not for the elements that can be obtained from them. Examples of this type of minerals are clays such as kaolin, and others like gypsum, magnesite, fluorite, etc.

Human beings increasingly use a greater variety of elements on the periodic table, and this is reflected in the graphs of historical extraction of mineral resources. If we analyse the evolution of the extraction of mineral resources worldwide (Fig. 2.5), we can see that in the last decades it has been growing continuously, which is also linked to the increasing variety of minerals that are extracted.

Non-metallic minerals, such as gypsum, limestone or phosphate rock, have experienced great growth, since these materials are widely used in the construction sectors

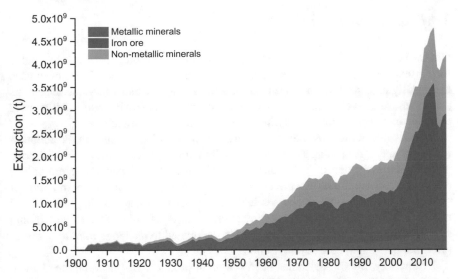

Fig. 2.5 Primary extraction of mineral resources from 1900 to 2018. The effect on the industry of the 2008 crisis and its subsequent recovery can clearly be observed. *Data source* USGS

and, in the specific case of phosphate rock, in the food sector as a valuable source of phosphorus.

In Fig. 2.5, we can observe how iron, separated from the rest of the metallic elements due to its importance in weight, has seen its extraction multiplied drastically. Specifically, a maximum peak of iron extraction can be seen between 2010 and 2014, a period in which almost three billion tons were extracted per year, later experiencing a drop in 2015 (Stanek et al., 2017). A similar situation can be observed, although on a smaller scale, in other metallic elements such as chromium or titanium.

2.3 Some Strategic Minerals for the Present and the Future

Throughout history, there have been a series of elements that have always been of particular interest to human beings. This is the case of gold, a fundamental element not only for its economic value but also aesthetic value, as we have seen for the mine of *Las Médulas* in León (Spain). Furthermore, in the last two decades, other elements that until now had not been widely used have seen how the appearance of new applications in technological sectors has increased interest in their exploration and extraction. For this reason, gallium or tellurium, which were not broadly known and whose applications were much smaller until the last few decades, are starting to sound more familiar. Let's take a look at some of them.

2.3.1 A Classic: Gold

Gold, whose chemical symbol is Au, has been considered one of the most valuable metals since ancient times, appreciated for more than 6,000 years. It has been used historically in jewellery, for minting coins and as a safe investment value. Due to its properties, such as electrical conductivity, resistance to corrosion and stability, gold also has many applications today in electronics, alloys, and medicine.

It is also one of the most expensive metals in the world, after some of the elements in the platinum group such as rhodium or palladium. Figure 2.6 shows the real price and the nominal price of gold since the beginning of the twentieth century. The difference between these two prices is that the nominal price considers the price of gold at the time it was produced, while the real price is adjusted taking into account inflation, in this case from 1998. In any case, the price of gold has been undergoing several significant changes and these peaks are related to a change in mining trends (Wellmer and Calvo et al., 2018; Scholz, 2017).

Historically, South Africa has been the main producer of this valuable element for many decades, although it has fallen several places in the ranking in recent years. Specifically, in 2017 around 3,200 tons of gold were mined globally, not counting illegal mining, almost 40% corresponding to China, Australia, Russia and the United States. The rest of the extraction was distributed across some 80 countries, with South Africa relegated to sixth place (USGS, 2019). In the case of Spain, during that year only 1.6 tons of gold were mined, a considerably small figure if we consider the importance that it came to have worldwide in the past.

Of the more than 4,000 tons of gold demanded in 2017, around 53% globally went to jewellery, and 39% to investments (Fig. 2.7). Much of the investment gold was purchased by banks out of concern about slowing global growth or rising geopolitical

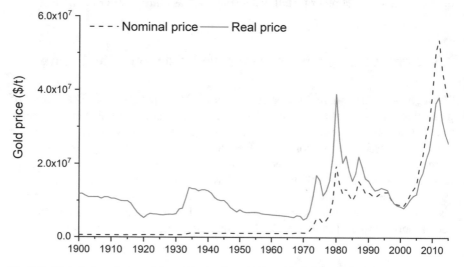

Fig. 2.6 Historical evolution of the gold price from 1900 to 2015. *Data source* USGS

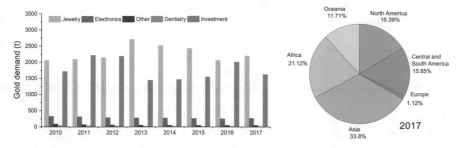

Fig. 2.7 Gold demand by sector (left) and gold extraction by region during 2017 (right). *Data sources* USGS

tensions. According to September 2019 data from the World Gold Council, the gold market development organisation, the United States is the country whose banks have the most gold reserves, with more than 8,400 tons, followed by Germany (3,300 tons) and Italy (2,400 tons). Spain has only about 280 tons of gold in its banks' reserves.

As stated before, we can differentiate the concepts of "extraction" and "production", especially when talking about gold since, although they may seem similar, their implications are different. By extraction we refer to material originating from mines; production can be primary or secondary; primary if it originates from mines, secondary if it originates from reprocessing or recycling. However, gold used in electronic components is rarely recovered due to the lack of adequate processes and technologies, and much of the jewellery is kept at home for its sentimental value and is rarely recycled.

2.3.2 Rare Earth and Other Essential Elements

Rare earth elements are a set of 17 elements of the periodic table composed of the 15 lanthanides plus yttrium (Y) and scandium (Sc), whose applications range from catalysts to ceramic and metallurgical uses.

Several essential elements for the development of new technologies and renewable energies belong to this group of rare earth elements, such as neodymium (Nd) and dysprosium (Dy) used in permanent magnets in wind energy. In addition, we find an increasing number and variety of rare earth elements in efficient lighting (fluorescent and LEDs), vehicles or electronics. It is not surprising, therefore, that their production has multiplied by seven in the last 40 years. These elements are essentially extracted from two minerals, monazite and bastnaesite and, despite their name, they are not especially rare in nature. Some rare earth elements are even as abundant as lead, but they are so named because they usually appear in a very dispersed form. They are always mixed with each other, and technologically, the most critical and energy-consuming stage is their separation.

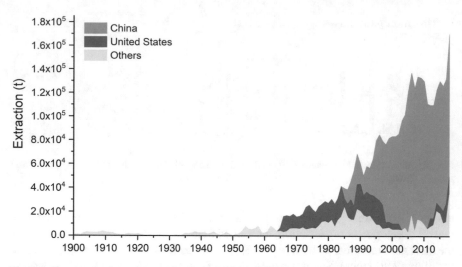

Fig. 2.8 Rare earth element extraction from 1900 to 2018 by country. *Data sources* USGS

In the first half of the twentieth century, the extraction of these elements was almost insignificant, experiencing a great boom in its second half (Fig. 2.8). In the 1950s extraction was only around a few thousand tons and the main source was placer deposits, which will be explained later in the book.

In the mid-1960s, rare earth elements were also mined in the United States, specifically at the Mountain Pass mine, located in the state of California. In 1985, China also began to extract rare earth elements, with Bayan Obo among the most important mines, reaching a total monopoly at that time. China's entry into the rare earth market not only led to an increase in global extraction, but also to a sharp drop in price in the early 1990s, endangering the continuity of other mines.

There is a certain amount of rare earth deposits scattered around the world. Although due to their properties, rare earth elements are sometimes accompanied by some radioactive elements such as thorium or uranium, which can cause many problems during extraction. An example of this type of environmental problem was seen in Bukit Merah, in Malaysia, when a rare earth refinery contaminated the waters and soils of the area with radioactive waste (Ali, 2014).

The Mountain Pass mine closed in 2002, after a series of accidents that took place in 1998, precisely related to radioactive waste dumping in Lake Ivanpah. However, due to price increases of rare earth elements, changes of ownership and decontamination studies (among others), it resumed its activity in 2012, restoring the country to its historical position as the world's supplier of these elements.

Currently, globally, the extraction of rare earth elements is 140,000 to 170,000 tons per year, although the data is not entirely accurate due to the lack of specific figures on artisanal extraction in some parts of China. Despite this, from the information available, China maintains its position as the world's leading producer of rare earth elements, being responsible for almost 60% of the extraction in 2019 (USGS, 2020).

This monopoly has already caused certain problems and incidents in the recent past. In October 2010, China limited rare earth exports to Japan after a dispute over a Chinese fishing vessel. In September of that year, the Japanese coast guard intercepted a fishing boat that had allegedly been fishing illegally in the Japanese-claimed sea zone surrounding the Senkaku islands. The ship's captain was arrested after showing resistance but was released days later due to pressure from China. Despite not having explicitly promoted, an order to deny exports of rare earth elements to Japan customs procedures and inspections became so difficult that finally, in October, rare earth elements stopped arriving on the Japanese island, a situation that was not resolved until November (Neill & Speed, 2012). Japan is one of the main receivers of rare earth elements mined in China, and many countries rely on magnets and other items manufactured in Japan that use these elements. The crisis of exports from China in 2010, where they decreased to 86% in just one year, caused great uncertainty in the market (Fernandez, 2017). The price of rare earth elements increased threefold from 2010 to 2011, which triggered a boom in research and exploration in other parts of the globe. This specific conflict also directly affected rare earth elements bought by the United States and Europe, thus highlighting the problems of the situation and the possible use of these elements as a political weapon. Why export rare earth elements when they can export wind turbines whose added value is much greater? Newspapers like the Wall Street Journal or the Financial Times had already alerted the world that by controlling the key to exports, China will come to control the global development of new sustainable technologies.

It must be considered that the mining sector has a very slow response time both to the increase in consumption and to the increase in prices. If an item's price suddenly drops, it is easy to stop mining or stop a mine, but if it increases, the time required to open a new mine can exceed ten years in the vast majority of cases. Following this rare earth price crisis, in 2013 it returned to previous values, again slowing interest in exploration. In recent years, some countries have been attempting to compete with China, to avoid a monopoly, such as the United States and Australia, reopening mines and increasing exploration efforts.

2.3.3 Technological Materials: Cobalt, Lithium, Niobium and Tantalum

Two elements whose names are now becoming familiar in recent years are lithium (Li) and cobalt (Co). Both elements are essential for the development of electric batteries. In 20 years, cobalt and lithium's primary extraction has increased five-fold and eightfold, respectively (Kelly & Matos, 2016). In 2017, the price of cobalt doubled in less than a year and has already become one of the most critical elements in the automotive sector, which is strongly betting on electric vehicles.

Lithium is obtained from two main sources: from the Chilean salt flats and from spodumene, a mineral that is mainly extracted and exported from Australia. Although worrisome, lithium does not appear to present as many supply problems as cobalt, which is found mostly in the Democratic Republic of the Congo. Specifically, the case of lithium will be taken up later to discuss current and future extraction trends, as well as possible associated supply risks.

Widely known is also the case of *coltan*, a black metallic mineral made up of a mixture of two minerals that are also highly valued, columbite and tantalite. These two minerals contain niobium, also called columbium (Nb), and tantalum (Ta), and their extraction is concentrated, as is the case with cobalt, in the Democratic Republic of the Congo (Fig. 2.9). Tantalum is used almost entirely in electronic devices, specifically, in capacitors and also in different parts of smartphones such as lenses, batteries, microprocessors, etc., whereas niobium, in addition to replacing and complementing tantalum, is used in aviation turbines and other materials with very high performance.

The presence of these strategic and much-needed elements in a politically and socially unstable country is the source of numerous conflicts and power struggles. These conflicts also directly affect border countries such as Rwanda, not to mention the miners' deplorable working conditions and the precarious—or indeed non-existent—safety and environmental protection measures. Many of the miners are children, who have been known to extract the mineral with their own bare hands (Amnesty International, 2016). That being said, different political, social and environmental conflicts related to lithium extraction are also emerging in the main producing countries.

Fig. 2.9 Luwowo coltan mine, located near Rubaya (the Democratic Republic of the Congo). Luwowo is one of the mines in this country that is certified by the International Conference of the Great Lakes (ICGLR) that ensures that the extracted coltan comes from conflict-free mines and that it meets minimum social standards. *Author* MONUSCO Photos. CC BY-SA 2.0. Wikimedia Commons

2.3.4 Indium, Gallium and Tellurium: The New Horizons of Photovoltaics

Indium (In), gallium (Ga) and tellurium (Te) are three other key elements for new technologies and specifically for photovoltaics, in addition to silver (Ag). Every year, the production of these elements together barely exceeds 1,000 tons and yet they are becoming more common in our society and even in our homes.

They are all obtained as by-products from the processing of other minerals and rocks that have small amounts of these elements in their structure. Consequently, referring to indium, gallium or tellurium extraction is uncommon. Instead, there is an increasing tendency to refer to production in refineries, also implying that global estimates are somewhat imprecise.

In zinc and copper mineral refining plants, it is common to recover indium in the slime generated during the process, along with many other elements present in small quantities, which are subsequently recovered and sold. For example, Bolivia is a country where zinc is one of the most important minerals, both in volume and in exports, and it is in these zinc concentrates where indium appears in not inconsiderable concentrations. Bolivia is the third country in terms of production of indium from mining concentrates, but rather than refining it, it merely exports it to other countries where this recovery is carried out (Zapata Rosso, 2018).

Furthermore, gallium is produced mainly from the processing of bauxite, a type of rock that has high aluminium content and gallium concentrations of the order of 50 parts per million (about 50 g of gallium per ton of bauxite). In the case of tellurium, it is largely recovered from the copper industry, from the anode slimes that are generated when refining the metal.

Indium and gallium are used in thin-film photovoltaic technology called CIGS (copper-indium-gallium-selenium), while tellurium is used in cadmium telluride (CdTe) cells. Both technologies offer the best performance in terms of market efficiency and that is why their production will increase in the coming years. As we have seen, the problem is that they depend on the use of elements that are very scarce in the Earth's crust.

Besides, gallium, indium and tellurium have applications in many other technological fields. For example, touchscreen technology requires ITO (indium tin oxide), a semiconductor material present in smartphones or electronic tablets. Indium and gallium are also present in LED technologies, giving different colours to efficient lighting.

2.3.5 Phosphorus: The Next Green Gold

Phosphorus (P) is much better known than the previous elements because it is an essential nutrient in agriculture: along with nitrogen (N) and potassium (K), it is the basic ingredient of fertilizers. Plants absorb these nutrients from the soil, which are naturally present, but sometimes insufficient to supply the crops demand. For this reason, fertilizers are and will be crucial in the future. With an increase in population, even more yield will be required from the soil to be able to feed the world population.

While potassium and nitrogen are abundant in nature, phosphate rock, from which phosphorus is extracted, is more concentrated and can be profitably mined from just a few countries. This phosphate rock can be made up of different phosphorous minerals, but one of the most common is apatite, a calcium phosphate. In Western Sahara we find the highest concentrations of phosphate rock worldwide, although this mineral wealth is also a source of social and political conflicts.

Nutrients, unlike the rest of the elements, cannot be substituted. It is not possible to feed a plant with aluminium or copper when phosphorus is scarce. A sector in direct competition with agriculture is that of biofuels, considered the fuels of the future, derived from biomass that also needs soil and nutrients to grow. This means that phosphorus may well become the next "green gold".

Perhaps with this example, we can now better understand the reason behind the conflicts in some inhospitable areas of the planet. And yet the opportunities for phosphorus recycling are immense since all living things are made of phosphorus (present in bones and teeth) and phosphorus is contained in the urine and excretions of all animals.

2.4 Mineral Criticality

Whether we are talking about fossil fuels or minerals, global extraction has only increased and, as we have seen in Figs. 2.1 and 2.5, it follows a general exponential trend. With this increase in demand, problems related to the availability and scarcity of certain resources have begun to be exposed, in particular in those sectors where a shortage of materials, such as the renewable energy sector, could put the global economy in check.

The alternative to fossil fuels, currently the main sources of energy, is the use of renewable energy sources such as the Sun, wind or water. However, in the case of minerals, the situation is relatively more complex due to their limited geographic distribution and not so obvious substitutability.

For this reason, and in order to avoid dependence on third countries, different classifications and lists of minerals considered "critical" have appeared, which vary significantly depending on the region, the criteria used and the objective sought. Before we look at existing criticality studies, though, it is important to define what a critical natural resource is. According to some authors, a natural resource can be considered critical when it is scarce and at the same time it is fundamental for the modern economy (van Oers & Guinée, 2016). Well-known examples of this type of resource are rare earth elements, already mentioned, which are mined in very few areas globally and whose use in new technologies has only increased.

Globally, the United States has a long tradition of analysing materials that are critical to its economy and security. The Defense National Stockpile Center was created after World War II to address future shortfalls in material availability, especially for the military industry. This centre is responsible for purchasing and accumulating specific quantities of certain elements, fundamental for the country's military, industrial or civil development, thus ensuring future supply and avoiding possible supply crises. For example, having a continuous supply source of tungsten played a fundamental role in World War II as a continuous supply of this mineral, very resistant to high temperatures, allowed the production of high-performance steels to manufacture weapons and ammunition (Calvo et al., 2019). This centre monitors more than 160 minerals of which 92 meet at least one of the several requirements to be considered vulnerable (U.S. Department of Defense, 2015).

In 2014, U.S. Congress allocated approximately \$41 M to purchase sufficient quantities of six products, including ferro-niobium, which is used as an alloying agent for stainless steel and steels resistant to high temperatures. Besides, in its 2015 report, it recommended accumulating amounts of 12 other resources, including antimony, beryllium (in the form of metal), germanium, lanthanum or europium (U.S. Department of Defense, 2015). In the specific case of antimony, an element used in various applications in the arms industry (missiles, ammunition, artillery shells, fire-retardant fabrics, etc.), the United States imports 85% of the amount it needs from China and Mexico, and it does not have mines in the country or refineries in which to obtain it.

With Donald Trump's arrival in the White House in late 2017, the President issued an Executive Order called "Federal Strategy to Ensure a Safe and Reliable Supply of Critical Minerals" (U.S. Department of Commerce, 2019). A year later, a list of 35 critical minerals for the country was published, as shown in Table 2.1. The criteria followed for drawing up this list are based on two factors; firstly, appearance in the Hirfindal-Hirschmann index, which measures the concentration of extraction in the countries. Second, the dependence on imports from the United States, based on the annual reports prepared by the United States Geological Survey (USGS).

Other countries, such as Korea, Japan or the United Kingdom, historically dependent on the supply of minerals, have also developed strategies and lists of elements that are critical to their respective economies (Bae, 2000; British Geological Survey, 2015; Department of Industry, 2019; Kawamoto, 2008). Even a country with as many mineral resources as Australia, where two-thirds of exports correspond to mineral and energy resources, has a policy regarding strategic minerals, although the approach is slightly different. The objective is to analyse which mines in the country have the potential to produce critical minerals in the future, such as southern Australia's Olympic Dam mine, which as well as containing deposits of copper, uranium, gold and silver, also contains significant amounts of rare earth elements that are not processed and that end up in dumps (Mudd et al., 2018).

In the case of the European Union (EU), as early as 1975 concern about the supply of raw materials began. Since then, the increasing costs of energy and the high dependence of this region on imports of certain resources have been hot topics on the political agenda. On the one hand, the EU has quite a few deposits of raw materials. However, in terms of exploration and extraction, there is competition with other land uses and environmental and technological limitations. On the other hand,

Table 2.1 List of critical minerals from different countries and main producing country and extraction percentage of each

	Critical mineral	United States[a]	Japan[b]	European Union				Main producing country[g]	Extraction percentage[g]
				2020[c]	2017[d]	2014[e]	2010[f]		
1	Aluminium (bauxite)	x		x				Australia	25
2	Antimony	x	x	x	x	x	x	China	71
3	Arsenic	x						China	69
4	Baryte	x		x	x			China	34
5	Beryllium	x		x	x	x	x	United States	74
6	Bismuth	x		x	x			China	81
7	Boron			x	x	x		Turkey	NA*
8	Caesium	x						NA*	NA*
9	Chrome	x	x			x		South Africa	44
10	Cobalt	x	x	x	x	x	x	DR of Congo	64
11	Coking coal			x		x		NA*	NA*
12	Fluorspar	x		x	x	x	x	China	60
13	Gallium	x	x	x	x	x	x	China	95
14	Germanium	x	x	x	x	x	x	China	63
15	Graphite (natural)	x	x	x	x	x	x	China	68

(continued)

Table 2.1 (continued)

	Critical mineral	United States[a]	Japan[b]	European Union				Main producing country[g]	Extraction percentage[g]
				2020[c]	2017[d]	2014[e]	2010[f]		
16	Hafnium	x		x	x			NA*	NA*
17	Helium	x			x			United States	40
18	Indium	x	x	x	x	x	x	China	40
19	Lithium	x	x	x				Australia	60
20	Magnesium	x	x	x	x	x	x	China	66
21	Manganese	x	x					South Africa	31
22	Niobium	x	x	x	x	x	x	Brazil	88
23	PGM[h]	x	x	x	x	x	x	South Africa	70
24	Phosphate rock			x	x	x		China	52
25	Potash	x						Canada	29
26	REE	x	x	x	x	x	x	China	71
27	Rhenium	x	x					Chile	55
28	Rubidium	x						NA*	NA*
29	Scandium	x		x	x	x		NA*	NA*
30	Silicon (metal)			x	x	x		China	60

(continued)

Table 2.1 (continued)

	Critical mineral	United States[a]	Japan[b]	European Union				Main producing country[g]	Extraction percentage[g]
				2020[c]	2017[d]	2014[e]	2010[f]		
31	Strontium	x			x			Spain	38
32	Tantalum	x	x	x	x		x	DR of Congo	39
33	Tellurium	x						China	68
34	Tin	x						China	29
35	Titanium	x	x	x				Canada	14
36	Tungsten	x	x	x	x	x	x	China	81
37	Uranium	x						Kazakhstan	39
38	Vanadium	x	x	x	x			China	55
39	Zirconium	x	x					Australia	39

* Data not available (NA)
[a]U.S. Department of Commerce (2019)
[b]Department of Industry (2019)
[c]European Commission (2020)
[d]European Commission (2017)
[e]European Commission (2014)
[f]European Commission (2010)
[g]USGS (2019)
[h]PGM: Platinum Group Metals

it largely depends on the external supply of critical raw materials, such as antimony or rare earth elements, which come mainly from China, as we have seen, or the case of niobium, which comes from Brazil and DR Congo.

In 2008, the *Raw Materials Initiative* (RMI) was created, tasked with preparing periodic reports on identifying raw materials that are critical to the region and their impact on the industry, among other things. One of the objectives sought by this initiative is establishing a common strategy to respond to the different challenges related to access to raw materials for all countries belonging to the European Union. This initiative is based on three fundamental pillars (European Commission, 2008). The first pillar is to ensure a fair and sustainable supply of raw materials from global markets, promoting greater international cooperation, managing trade and regulatory policy, promoting investment, etc. The second, promoting a sustainable supply of raw materials from European sources. For this, the investigation of mineral deposits in each country is encouraged to improve the existing knowledge base, including joint research projects among various members of the community, etc. Finally, the third pillar is based on reducing the consumption of raw materials and increasing resources, fostering recycling, the substitution of materials or increasing the use of secondary raw materials, among other measures.

Another objective of the Raw Materials Initiative is publishing and updating a list of raw materials considered critical to the European Union, whose evolution can be seen in Table 2.1. The first report was published in 2010, where 41 materials were analysed, of which 14 were considered critical (European Commission, 2010). In 2014, that initial list was expanded to a total of 20, of which 13 were already in the previous report (European Commission, 2014). In 2017, a new update was carried out, considering critical a total of 27 raw materials (European Commission, 2017). The most recent report is from 2020, the list was expanded to a total of 30 critical raw materials (European Commission, 2020). Compared to the previous report, chromium, coke and magnesite have been removed as critical raw materials. Instead, others, such as barium, hafnium, scandium, tantalum, and vanadium have been added (Fig. 2.10).

The methodology used by the European Union to determine if a material is critical is based on two factors: economic importance and supply risk. In the first case, the economic importance is calculated considering the primary extraction of each material and the sectors in which it is used, associating it with the GDP of each sector. Supply risk is divided into two categories, supply risk linked to political instability in producer countries and environmental risks linked to their corresponding environmental standards. Thus, among the factors used to calculate each element's risk of supply are the world governance index (WGI), the exchange of materials and the origin of European imports, the possibility of substituting one element for another and recycling rates. These last two factors can reduce the risk of supply. Therefore, starting from minimum thresholds for the two main factors—economic importance and risk of supply—we can obtain the definitive list of critical materials.

All of these criticality reports are focused on specific countries or regions and, although the importance of materials is taken into account in some way in the different sectors in which they are used, there are disruptive technologies that could penetrate

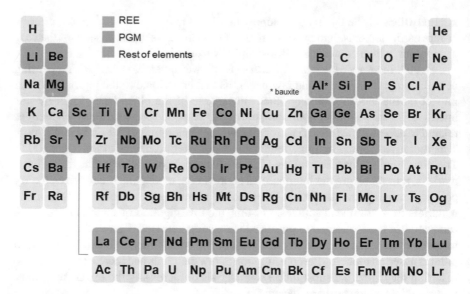

Fig. 2.10 Periodic table where elements considered critical to the European Union in the 2020 report are highlighted

in the short and the medium term. For instance, the current boom in renewable energy is causing an increase in demand for certain materials, some of which are critical. This in turn leads to doubts over future availability to cover this demand. This is an issue of vital importance for the energy and environmental transition and will therefore be given particular attention in Chap. 6.

Furthermore, all the criticality studies that we have seen previously converge on the same ideas: the existence of monopolistic and underdeveloped markets, a rapid growth in demand in certain sectors, a shortage of supply to cover this demand and fluctuations in prices. To these points, we must add tensions in political and international relationships, the existence of regulatory barriers, unequal geological distribution, environmental mismanagement of mining in certain countries, and a lack of research and new mining investments.

The problem of the growing demand for mineral resources and fossil fuels is increasingly pressing: every year the world's population increases by millions and our demand for materials continues to rise to meet our needs and manufacture different products. We can only source these raw materials from our planet, and they exist in limited quantities. Will there be enough minerals on the planet to supply this growing demand? What are the impacts associated with this extraction? We will analyse these issues in the next chapter.

References

Ali, S. H. (2014). Social and environmental impact of the rare earth industries. *Resources, 3*, 123–134.

Amnesty International. (2016). *This is what we die for. Human rights abuses in the democratic Republic of the Congo power the global trade in cobalt.*

Bae, J. C. (2000). *Strategies and perspectives for securing rare metals in Korea. Ritical elements for new energy technologies, an MIT energy initiative workshop report.*

BP. (2019). *British petroleum. Statistical review of world energy.* Available at: https://www.bp.com/content/dam/bp/business-sites/en/global/corporate/pdfs/energy-economics/statistical-review/bp-stats-review-2019-full-report.pdf.

British Geological Survey. (2015). *Risk list 2015.* Available at: http://www.bgs.ac.uk/mineralsuk/statistics/risklist.html.

Calvo, G., Valero, A., & Valero, A. (2018). Unfortunately, the amount of gold on earth is not infinite, a response to Wellmer and Scholz (2017). *Resources, Conservation and Recycling, 133,* 155–156. https://doi.org/10.1016/j.resconrec.2018.02.018.

Calvo, G., Valero, A., & Valero, A. (2019). How can strategic metals drive the economy? Tungsten and tin production in Spain during periods of war. *The Extractive Industries and Society, 6,* 8–14.

Department of Industry Innovation and Science. (2019). *Australia's critical minerals strategy. Australian Goverment.* Available at: https://www.industry.gov.au/sites/default/files/2019-03/australias-critical-minerals-strategy-2019.pdf.

European Commission. (2008). The raw materials initiative—meeting our critical needs for growth and jobs in Europe. Communication from the Commission to the European Parliamente and the Council. COM (2008) 699 final.

European Commission. (2010). *Critical raw materials for the EU. Report of the Ad-hoc working group on defining critical raw materials.*

European Commission. (2014). *Report on critical raw materials for the EU. Report of the Ad hoc working group on defining critical raw materials.*

European Commission. (2017). *Study on the review of the list of critical raw materials. Critical raw materials factsheets.*

European Commission. (2020). *Study on the EU's list of critical raw materials (2020).* Final Report. https://doi.org/10.2873/904613.

Fernandez, V. (2017). Rare-earth elements market: A historical and financial perspective. *Resources Policy Pergamon, 53,* 26–45. https://doi.org/10.1016/J.RESOURPOL.2017.05.010.

Kawamoto, H. (2008). Japan's policies to be adopted on rare metal resources. *Science & Technologie Trends, 27,* 57–76.

Kelly, T. D., & Matos, G. R. (2016). *Historical statistics for mineral and material commodities in the United States: U.S. Geological Survey Data Series 140.* Available at: https://www.usgs.gov/centers/nmic/historical-statistics-mineral-and-material-commodities-united-states.

Mudd, G., et al. (2018). *Critical minerals in Australia: A review of opportunities and research needs.* Record 2018/51. Geoscience Australia, Canberra.

Neill, D. A., & Speed, E. (2012). *The strategic implications of China's dominance of the global rare earth elements (REE) market. Defence R&D Canada. Centre for Operational Research and Analysis.*

van Oers, L., & Guinée, J. (2016). The abiotic depletion potential: Background, updates, and future. *Resources, 5*(16). https://doi.org/10.3390/resources5010016.

Stanek, W., Valero, A., Calvo, G., & Czarnowska, L. (2017). Resources. Production. Depletion. In: Stanek W. (Eds.) *Thermodynamics for Sustainable Management of Natural Resources.* Green Energy and Technology. Springer, Cham. https://doi.org/10.1007/978-3-319-48649-9_2.

U.S. Department of Commerce. (2019). A federal strategy to ensure secure and reliable supplies of critical minerals. Available at: https://www.commerce.gov/data-and-reports/reports/2019/06/federal-strategy-ensure-secure-and-reliable-supplies-critical-minerals.

U.S. Department of Defense. (2015). *Strategic and critical materials. 2015 Report on stockpile requirements.* Available at: https://www.hsdl.org/?view&did=764766.

U.S. Energy Information Ad. (2019). *EIA adds new play production data to shale gas and tight oil reports.* https://www.eia.gov/todayinenergy/detail.php?id=38372.

USGS. (2019). *Mineral commodity summaries 2019. United States geological service.* Available at: https://prd-wret.s3-us-west-2.amazonaws.com/assets/palladium/production/atoms/files/mcs2019_all.pdf.

USGS. (2020). *Mineral commodity summaries 2020. United States geological service.* Available at: https://pubs.usgs.gov/periodicals/mcs2020/mcs2020.pdf.

Wellmer, F.-W., & Scholz, R. W. (2017). Peak gold? Not yet! A response to Calvo et al. (2017). *Resources, Conservation and Recycling, 134,* 313–314. https://doi.org/10.1016/j.resconrec.2017.11.015.

Zapata Rosso, A. (2018). *El indio en Bolivia: muchos ceros a la izquierda. Reporte de Industrias Extractivas 4 (CEDLA).* https://doi.org/10.13140/RG.2.2.21737.19042.

Chapter 3
On the Availability of Resources on Earth

Abstract Fossil fuels and minerals are non-renewable resources. What's available on Earth to extract is finite and is determined by several factors such as technology, global resources and economic value. This section will first introduce the different terms used in the mining industry related to raw materials and how they are formed. Then, we will proceed to describe the mining process. In a mining project, up to four operations can be distinguished: exploration and research, development phase, operation, which includes processing and beneficiation, and last, closure and reclamation, which aims to return the land to the conditions before the existence of mining activity. All these phases depend on the type of mineral or raw material. Descriptions about pyro and hydrometallurgical copper processing will be used as examples to better understand how the minerals are extracted and converted into items that can be used in further steps of the value chain. Last, the impacts of the sector will be considered from an environmental to social perspective.

Earth formed some 4,500 million years ago, from the accretion of other smaller bodies. During this process a vast increase in temperature took place, eventually resulting in complete or partial melting; some of the planet's current inner heat is a remnant from that process. Since its formation, our planet has been cooling down due to a balance between the energy received from the Sun and the emission of heat from the Earth's surface. With the passage of time, this has been generating a series of geological processes that have originated the mineral deposits that we know today.

In this chapter, we will explain the main characteristics of energy and non-energy mineral deposits, their availability, how to exploit them to produce raw materials, as well as the environmental and social implications associated with their extraction.

A. Valero et al., *The Material Limits of Energy Transition: Thanatia*,
https://doi.org/10.1007/978-3-030-78533-8_3

3.1   Resource Classification

A natural resource can be defined as any form of energy or matter obtained from nature without the involvement of human beings. Air, water, biomass or minerals can be classified as natural resources. A distinction can be made between renewable and non-renewable natural resources, understanding the former as those that regenerate on a human time scale. Examples of this type of resource are solar radiation, hydroelectric energy, geothermal energy or wind. This category can also include wood and biomass if obtained in a sustainable manner. Non-renewable resources are those that do not regenerate at the same rate as they are being consumed. These can be subdivided into energy minerals, such as oil, coal, natural gas (generally called fossil fuels), as well as uranium, and non-energy minerals (hereinafter *minerals*), consisting of the remaining substances (iron, copper, aluminium, gypsum, salt, etc.).

When classifying the amounts of minerals or fossil fuels available on our planet, we use two terms depending on our degree of knowledge: reserves and resources. Different categories exist within each group. There are different classifications depending on the way the data is collected and how the analyses have been carried out in the field, which may vary between mining companies, geological services, etc. Several initiatives exist at a global level to standardise reporting to facilitate comparisons between them.

One of the best-known generic classifications is the so-called "McKelvey box", developed by the United States Geological Survey (Fig. 3.1). Using this tool as our basis, we will explain the definitions of reserves and resources. Resources are defined as a natural concentration of a solid, liquid or gas in the Earth's crust in such

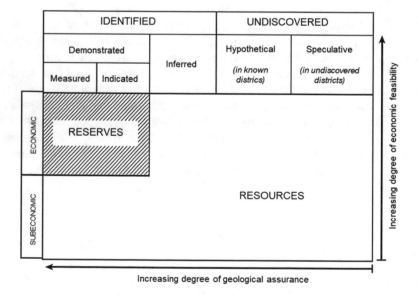

Fig. 3.1 McKelvey classification of reserves and resources. *Source* USGS

a form and quantity whose extraction is currently or potentially possible. Among the resources, we can distinguish those that are in already known mining districts, such as hypothetical resources, and those that are in districts that have not yet been explored, also called speculative resources, and that are usually obtained from 3D models. As such, resources are an extremely broad concept, encompassing mineral concentrations that may or may not be exploited.

The term reserves apply only to that part of the mineral deposits that can be recovered, obtaining an economic benefit, at the time the analysis is carried out. Therefore, reserves are a small part of the resources that meet a series of criteria whereby their extraction is profitable. Reserves, from a higher to a lower degree of geological knowledge, are subdivided into measured and indicated reserves, the latter having a lower degree of reliability.

Logically, there is no clear limit between reserves and resources, since the classification depends on the theoretical geological data obtained in the field, the price of each element, the degree of recoverability, etc. A change in the recovery technology or a decrease in an element price can change the economic classification of a deposit: what is not profitable to be extracted today may become so in 5 years, and vice versa. For example, the fact that it became economical in the last century to exploit porphyry copper deposits due to changes in technology increased copper reserves, and therefore resources, by over 100%.

For the reasons discussed above, reserves are not usually a good indicator for assessing the mineral wealth available on Earth due to the uncertainties and associated timing. In fact, the world's reserves of most minerals are now greater than in any past era due to greater geological information and the existence of more efficient technologies. The best approach to assess the existing mineral wealth would be to use data on resources. However, as this classification is much more extensive and often of a purely theoretical origin, the information available is usually scarce, imprecise and incomplete.

An example is gallium, an element that is extracted as a by-product of the processing of bauxite and zinc minerals. The information that exists on gallium resources is based on the theoretical average amount of gallium that bauxite contains—as mentioned before, about 50 g per ton. In this way, the amount of gallium that could be potentially recoverable is calculated from the resource data of this rock. This would give us a resource figure of one million tons of gallium (USGS, 2020), an estimate that may not be too close to reality.

In addition to the concepts of reserves and resources, it is also essential to be aware of the so-called cut-off grade, a term mostly used when talking about metallic elements. The cut-off grade represents the minimum degree of concentration of the mineral, a threshold above which its extraction becomes economically feasible. The material that is above this ore grade is then called ore and is taken to the treatment plant, while the material below this grade is called gangue or waste rock, and usually ends in tailings. This cut-off grade is usually measured in grams per ton or as a percentage of contained metal and changes continuously depending on the sale price of the mineral. On occasion, it is profitable to reprocess tailings for reasons of past and present profitability or to attempt to recover materials whose recovery was not technically possible before, a topic that will be covered in Sect. 8.6.

It should be remembered that when it comes to ore grade in the case of reserves and resources, or even in company or agency reports, this does not necessarily have to coincide with the just mentioned cut-off grade, since the latter depends purely on economic factors. Similarly, within a mine, there are other types of definitions of ore grade based on economic and technical parameters that do not need to be explored at this time.

3.2   Formation and Availability of Fossil Fuels

Fossil fuels are energy minerals made up of the remains of plants or animals that accumulated their energy from the Sun millions of years ago and constitute a reservoir of chemical energy. The main commercial types are coal, oil and natural gas.

Coal is a sedimentary rock of organic origin that was formed millions of years ago from existing vegetation. About 360 million years ago, during the Carboniferous period, the large ferns and trees existing in swampy areas, as they died, were deposited in the soil in layers. Little by little they were transformed, thanks to the action of fungi and bacteria in an oxygen-poor environment. Subsequently, these layers were covered with others, thus increasing the pressure and temperature they were subjected to, resulting in compaction and decreased moisture in the sediments. During this process, the sediment's carbon content progressively increased, giving rise to the different types of coal. Coal layers also formed in other time periods, but their formation during the Carboniferous period represents one of the most significant instances.

Coal is primarily made up of carbon, oxygen and hydrogen, but it may also contain traces of other elements such as sulphur. It can also be accompanied by inorganic matter that includes silicates, oxides and metal carbonates in varying amounts. It is classified according to its carbon content, or equivalently, its calorific value, which indicates the amount of energy per unit mass that it gives off when oxidised. From highest to the lowest calorific value, they can be ordered as follows: peat, lignite, subbituminous, bituminous and anthracite. Peat is the coal with the lowest carbon content, formed in the first stages of plant matter transformation; it usually has high humidity and many volatile elements. Lignite has a little more carbon, between 60 and 75%, but plant remains can still be seen; coal, which includes the bituminous range, already has a carbon content between 75 and 90% and a very high calorific value. Lastly, anthracite is the most evolved carbon of all, with up to 95% carbon, very few volatile elements and a very shiny black colour.

Oil and natural gas originated from the accumulation of marine plankton that became hydrocarbons after a fermentation process. The formation of these hydrocarbons takes place in two phases: first, biological, chemical and physical processes act to break down organic matter into kerogen, a precursor of natural gas and oil (Calaway & van Rensburg, 1982). The thermal alteration of kerogen marks the second phase to hydrocarbons as more recent sediments accumulate above. This generation of hydrocarbons occurs at a temperature around 50–60 °C and a depth of between 2 and 2.5 km. This process then continues until the sediments reach 200–250 °C and a depth of between 6 and 7 km. Oil formation predominates in the lower temperatures and natural gas in the higher ones. With time, these components can migrate from the source rock, the rock where they formed, through the subsoil. As they are in a liquid and gaseous state, they can migrate via pores and cracks until they reach an impermeable layer. This impermeable layer prevents their circulation, then is retained in the pores and holes of the so-called reservoirs.

There are numerous classifications of oil based on its composition, including that of the British standard BS2869:1998, which distinguishes between distilled, residual, kerosene and heavy fuel oils. Natural gas contains mainly methane, but there may also be small amounts of ethane, butane, propane, carbon dioxide, nitrogen, helium and hydrogen sulphide. As oil and natural gas are more efficient and relatively cleaner energy sources than coal, they have been replacing this fossil fuel since the 1950s.

In terms of fossil fuels, instead of talking about reserves and resources, the term proven reserves is used. Proven reserves are defined as the amount (estimated with reasonable certainty) that could be recovered from known deposits with current technology and operating conditions, based on geological and engineering analyses.

The Middle East can be singled out when talking about globally proven reserves of each of the three types of fossil fuels, with almost half of the world's oil reserves and a third of those of natural gas. As for coal, the total world proven reserves are close to one trillion tons. The regions with the most proven reserves are the Asia Pacific and North America (Fig. 3.2), while Australia has 14% of the total proven reserves and the United States has 24%. Proven reserves of oil and natural gas are much less than those of coal. At the end of 2018, they were estimated at 244 Gtoe of oil and 197 trillion cubic meters of gas. The distribution by regions is also different: the Middle East predominates in oil and the Middle East and the Commonwealth of Independent States (CIS) both predominate in natural gas.

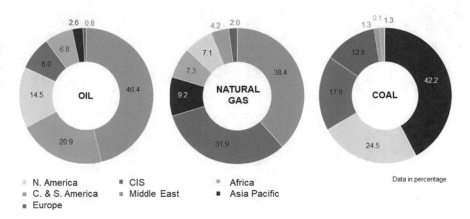

Fig. 3.2 Proven fossil fuel reserves by country (data in percentage). *Data sources* BP (2019)

Based on this availability, and assuming that the annual extraction of these fossil fuels in future years will be the same as in 2018, there will be enough coal for more than 250 years and natural gas and oil for about 50. In addition to conventional oil reserves, there is an additional amount of recoverable reserves between 38 and 141 Gtoe. With respect to additional natural gas resources, this amount could be between 200 and 500 Gtoe.

In exploration terms, the oil industry is relatively mature and the amount of additional reserves that could be discovered in the future is not clear (IPCC, 2000). It is widely believed that the number of oil fields being discovered is decreasing. Similarly, it is also believed that the increase in reserves is due to the revision of existing fields (underestimated thus far) rather than to new discoveries (Ivanhoe & Leckie, 1993; Laherrère, 1994; Campbell, 1997).

Figure 3.3 shows the evolution of the millions of barrels discovered in new oil and gas deposits from 2000 to 2017. Little by little, investment in exploration for oil and natural gas fields is recovering after having fallen almost 45% in recent years; although the volume of new discoveries has decreased slightly, it is still at the level of the 1950s (BP, 2018).

In 2018 and 2019, there have been several discoveries of new oil fields, particularly those of the Guyana coast, an area explored for several years by ExxonMobil. The United States Department of the Interior announced in 2019 the discovery of the largest continuous oil and gas field ever found, with a content of 46.3 trillion barrels of oil and 281 trillion cubic meters of natural gas (Gaswirth et al., 2018).

In addition to conventional fossil fuels, there are many unconventional fossil resources, including slates and oil sands. Bituminous slates are rocks with deposits of organic matter that have not been exposed to sufficient pressure and temperature to become oil, but that can be heated to generate liquid derived products. There are deposits in 33 countries, notably in the United States. Bituminous sands are made

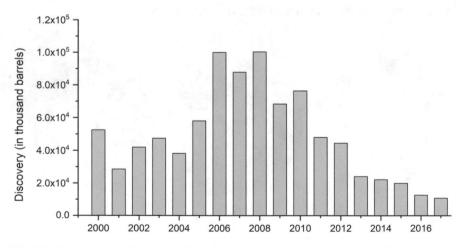

Fig. 3.3 Discoveries of oil and natural gas deposits in recent years. *Data source* BP

up of 83% sand and 10% bitumen (a mixture of black and highly viscous organic substances), as well as water and clays. The existence of deposits of this type is known in 23 countries, among which Canada predominates, from which 2.4 trillion barrels of oil could be extracted (International Energy Agency, 2018).

As mentioned, in conventional reservoirs, the natural gas expelled from the source rock migrates and is trapped when it reaches the impermeable layer, forming what is known as a reservoir. As for unconventional natural gas deposits, the gas is generated in the same way, the difference being that it does not migrate. Instead, it is retained in the pores of the rock or sands as they are very poorly permeable, and the extraction method is therefore completely different. In this case, these pores must be broken to release the contained natural gas, generating artificial cracks in the ground that serves as a path during its ascent.

Globally, unconventional fossil fuel reserves, such as slate and tar sands, among others, could quadruple those of conventional fossil fuels. However, they are more expensive to extract, so their profitability heavily depends on the price of a barrel of oil. For example, the slate refining process requires two to three times more energy than that used to produce conventional oil. To this, we must add the fact that the associated environmental footprint is much greater.

3.3 Formation and Availability of Non-energy Minerals

The physical and chemical properties of minerals are directly influenced by the two major energy sources on Earth: the Sun and the Earth's core. Both are responsible for the movement of materials from the Earth's core to the crust. Materials also move from the crust to seas or rivers through water currents, forming rocks in the so-called geochemical cycle. The resulting dynamic equilibrium is called the geochemical balance.

Minerals are not found in the same concentration throughout the crust nor do they all appear in the same way. The processes that form mineral deposits can be separated into four categories: magmatic, hydrothermal, metamorphic and surface processes, the first three being related to processes that take place inside the Earth's crust, and the last one on the surface. Regardless of the process, minerals tend to appear together in deposits in certain types of rocks due to their physical and chemical characteristics, known as a mineral association.

First, within the magmatic processes that give rise to mineral deposits, one can distinguish those in which the crystallisation of minerals occurs within the crust, as magma cools. High-temperature fluids can also lead to the formation of minerals, but these will be defined within the hydrothermal processes. The composition of the minerals will depend directly on the composition of the magma, which can be altered when it comes into contact with other rocks during its ascent to the surface.

Examples of deposits of magmatic origin are the deposits of chromium and minerals of the platinum group. One of the most representative mines where nickel, palladium and platinum are mined, in addition to non-negligible amounts of copper, cobalt and gold, is the Norilsk-Talnakh mine, located in Siberia. Interestingly, this mine alone is responsible for 1% of SO_2 emissions globally as it produces large amounts of this compound during the metallurgical process (Neukirchen & Ries, 2020). Another example is komatiite, a type of volcanic rock formed from extremely hot, fluid, magnesium-rich magma. Nickel and copper deposits in rocks of this type represent, worldwide, the largest nickel deposits and usually have higher concentrations than other types of deposits. A classic example of nickel and copper komatiite is located in Kambalda, Western Australia.

Secondly, hydrothermal processes are associated with the circulation of hot fluids through fractures and pores in the rocks, which migrate to more favourable places with lower pressures until equilibrium is reached. When the temperature decreases, when different fluids mix or when these fluids come into contact with certain elements and rocks, these fluids dissolve and the precipitation of new minerals occurs. The type of mineral that is formed will depend on various factors, notably the composition of the fluid, pH, temperature and Eh (redox oxidation potential).

This hydrothermal process can occur in almost any type of geological environment. The deposits it generates are very varied, serving as an example porphyry copper deposits, a type of mineralisation formed by magmatic and hydrothermal processes. They appear associated with magma intrusions that rise through cracks and cool near the surface. Its name, porphyry, is due to the typical association with igneous rocks that have a texture with large feldspar crystals surrounded by a very fine-grained matrix. They are usually very large deposits in terms of weight, from 1 million tons to several gigatons, but with very low element concentrations, so their extraction is usually done via open-pit mining rather than by underground mining.

Porphyry copper deposits are the main source of this metal worldwide, with more than 60% of copper coming from these deposits. Some of the best examples of porphyry copper deposits are found in Chile, in particular, Chuquicamata (Fig. 3.4), Escondida, Los Bronces or El Teniente (Bustillo Revuelta, 2018).

Fig. 3.4 Panoramic view of the Chuquicamata copper mine (Chile). Author: Diego Delso, delso.photo, License CC-BY-SA. Wikimedia Commons

Porphyry copper deposits can be categorised into several types according to the element that predominates in addition to copper: molybdenum, tungsten, tin or gold. In the latter case, a porphyry copper deposit is said to be copper and gold when the amount of gold present is greater than 30 ppm (Perello & Cabello, 1989). The best-known copper and gold porphyry deposits are in Papua New Guinea and Canada, although they also appear in Argentina, Australia, Brazil, the United States and Indonesia (Bustillo Revuelta, 2018).

Within the hydrothermal type deposits, generated by the presence of fluids at high temperatures, we also find massive sulphide deposits. They are small to medium in size, with high ore grades of base metals such as copper, zinc and lead, but significant concentrations of gold, silver, cadmium, selenium, bismuth, etc. can also be found. They are also one of the few types of deposits that can today be observed while they are forming since they are generated by the precipitation of sulphides and other elements on the ocean floor. This situation currently occurs in black fumaroles, columns through which a mixture of gases and vapours comes out at high temperatures where the minerals precipitate when they come into contact with cold water. These deposits have been known and have been exploited for more than 2,000 years, using the copper that was extracted, for example, from the famous mines of the Iberian Pyrite Belt for the manufacture of weapons during the Roman era (Bustillo Revuelta, 2018).

Although there are massive sulphide deposits distributed throughout practically the entire planet, the Iberian Pyrite Belt, located in Portugal and Spain, is one of the best known and most representative. It is an area with a length of about 250 km and a width of between 30 and 50 km containing an estimated 2,000 million tons of polymetallic sulphides, making it one of the highest known concentrations of this type today in the Earth's crust (Tornos Arroyo, 2008). One of the best-known mines is Rio Tinto, and there are currently several operating mines that extract copper, lead and zinc sulphides. Another well-known deposit, and that usually gives its name to this type of deposit, is Kuroko, in Japan.

Let us now return to the different types of deposits. During the metamorphic processes, existing rocks and minerals transform due to the increase in pressure and temperature to which the rock is subjected; this gives rise to the formation of new types of rocks. An example of this type of deposit is marble, produced by limestone metamorphism. Another example is graphite, a mineral of economic interest that is produced by the recrystallisation of the organic matter present in the rocks caused by an increase in pressure and temperature.

Finally, within the processes that occur on the surface of the Earth, we can distinguish those of weathering of pre-existing rocks, which can create sedimentary rocks such as sandstones and conglomerates, and those related to chemical precipitation, which create deposits of limestones or gypsum.

In the first case, the weathering and disintegration of the rocks can generate placer-type deposits, when particularly dense minerals are accumulated. Some examples are placer deposits of gold, zircon, tin, etc. The high resistance to weathering of these minerals and their durability make them concentrate gradually over time, while the

Fig. 3.5 Piles of salt in the Salar de Uyuni (Bolivia). Author: Luca Galuzzi—www.galuzzi.it. Wikimedia Commons

rest of the elements of the original rocks disappear. The minerals that may appear vary, therefore, depending on the type of primary rock.

In the second case, the deposits generated by chemical precipitation are, for example, those formed through evaporation processes of lakes that can give rise to large deposits of gypsum and salts, as can be seen in the Salar de Uyuni in Bolivia (Fig. 3.5) or the Salar de Atacama in Chile.

All these types of deposits, and the minerals that appear in them, are present in the first 10–100 km of the Earth's crust, that is, they are concentrated in less than 0.4% of the Earth's total mass. The average concentration of the elements in the crust is what controls the appearance of mineral deposits. Nine elements, called major elements, make up 99.5% of the crust: these are oxygen, silicon, aluminium, iron, calcium, magnesium, sodium and potassium in order of importance. The remaining 0.5% is made up of minor and trace elements. The major elements are sufficient to form the most common mineral deposits, but the minor elements must undergo an enrichment process to appear in the concentrations known today in mines.

Very little is known about the Earth's crust, where the natural mineral resources accessible to humans are concentrated, given that the costs of exploration and research are extremely high. Most of the deposits currently mined are close to the surface, but the crust has an average thickness of 40 km and the deepest open-pit mine is less than 1 km. The deepest underground mine is 3.5 km, although most do not exceed 2 km. The distribution of minerals in the crust is also uneven and irregular, not only from an element point of view but also geographically. In some cases, even though there are known deposits of an element in many countries, it is only profitable to extract it in a few places. Although we can know the concentrations of a certain element in a given part of the planet, we need to access information at a global level in order to truly understand its availability.

Thanks to an extensive, but still incomplete, record of existing mineral deposits, we have reserves and resources data for the main minerals that humans extract today. These amounts of reserves are associated with an ore grade, that is, the minimum content of the metal in the mineral that makes its extraction profitable; in other words, the concentration that makes it possible to pay the costs of extraction, treatment and marketing of the mineral.

If we take copper as an example (Fig. 3.6), we can see the considerable difference between extraction in 2018 and available reserves (USGS, 2019). However, if we analyse the accumulated extraction between 1900 and 2018, we can see that it is very similar to the reserves figures of the USGS. In other words, if copper extraction remained constant, the available reserves could be depleted in 40 years. It must be remembered, however, that reserves represent the amount of copper that can be recovered with an economic benefit, at the time of analysis. In other words, this figure can vary considerably depending on many factors other than availability. Furthermore, assuming the same annual consumption as that of 2018, considering known resources, theoretically, there would be no issues with copper availability for almost 150 years. However, as mentioned, not all these resources are available or could be technically extracted, so this figure should be taken with caution.

The most abundant deposits in the crust are those of iron, followed by phosphate rock, potassium, manganese and aluminium. In contrast, deposits of indium, tellurium and PGM (platinum group metals) are the scarcest in the world. In the case of some industrial minerals, such as limestone or gypsum, the amount of reserves and resources is vast, so there is no point in talking about their reserves or resources.

Table 3.1 shows the world's reserves and resources of the main non-energy minerals of economic importance, according to the databases of the United States Geological Survey (USGS), for two different years, 1995 and 2018.

As we can see, reserves and resource data can be very dynamic. They change over time according to different factors. Figures can decrease as ore is mined or increase if

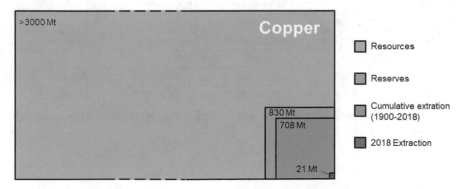

Fig. 3.6 Extraction data (2018), accumulated extraction (1900–2018), known copper reserves and resources. *Data sources* USGS (2019)

Table 3.1 World reserves and resources data for the main non-energy minerals

	Reserves (t) 1995	Reserves (t) 2018	Reserves (t) 1995	Reserves (t) 2018
Aluminium (bauxite)	23,000,000,000	30,000,000,000	75,000,000,000	75,000,000,000
Antimony	4,200,000	1,500,000	5,100,000	NA
Baryte	170,000,000	320,000,000	2,000,000,000	2,000,000,000
Beryllium	NA	NA	66	100
Cadmium	540	NA	NA	NA
Chromium	3,700,000,000	560,000,000	11,000,000,000	12,000,000,000
Cobalt	4,000,000	6,900,000	11,000,000	25,000,000
Copper	310,000,000	830,000,000	2,300,000,000	2,100,000,000
Fluorspar	210,000,000	310,000,000	330,000,000	5,000,000,000
Gallium	NA	NA	1,000,000	1,000,000
Gold	44	54	75	NA
Graphite	21,000,000	300,000,000	NA	800,000,000
Iron	150,000,000,000	170,000,000,000	800,000,000,000	800,000,000,000
Lead	68,000,000	83,000,000	1,500,000,000	2,000,000,000
Lithium	2,200,000	14,000,000	12,000,000	62,000,000
Manganese	680,000,000	760,000,000	NA	NA
Molybdenum	5,500,000	17,000,000	12,000,000	20,000,000
Nickel	47,000,000	89,000,000	130,000,000	130,000,000
Niobium	3,500,000	9,100,000	NA	NA
PGM	56	69	100	100
Phosphate rock	11,000,000,000	70,000,000,000	NA	300,000,000,000
REE	100,000,000	120,000,000	NA	NA
Silver	280	560	NA	NA
Tantalum	22	110	NA	NA
Tellurium	20	31	NA	NA
Tin	7,000,000	4,700,000	NA	NA
Titanium	300,000,000	940,000,000	230,000,000	2,000,000,000
Tungsten	2,100,000	3.300,00	NA	NA
Vanadium	10,000,000	20,000,000	63,000,000	63,000,000
Zinc	140,000,000	230,000,000	1,800,000,000	1,900,000,000

Data is provided for years 1995 and 2018, in order to appreciate the evolution. Data from the USGS.
NA = data not available

new mineral deposits are discovered, or if it becomes possible to economically mine deposits with very low ore grade due to technological changes. These numbers can also change in response to shifts in demand and due to political and socio-economic factors. Specifically, mineral reserves are highly influenced by the market price of the elements, and they also depend on the investments of mining companies. For example, the reserves of phosphate rock have multiplied almost sevenfold in the last 20 years; however, the reserves of other minerals, such as chromium or manganese, have decreased despite the increase in their price (Calvo et al., 2017).

Taking as an example the evolution of nickel reserves from 1995 to the present, we can clearly see that the data follows an upward trend (from 47 to 89 million tons) despite experiencing peaks and troughs. Nonetheless, in the case of chromium, the trend is precisely the opposite, going from almost 3.7 billion tons of reserves in 1995 to 560 million tons in 2018. In the case of resources, the figure for nickel has not changed at all, remaining fixed at 130 million tons. However, chromium resources did experience a slight increase in 2003, when they went from 11 to 12 trillion tons, a figure that has remained relatively unchanged to this day.

3.4 Mineral Extraction and Processing

In order to better understand the problems associated with mineral extraction, in this section, we will describe all the stages that are carried out in the mining industry. However, we must consider that, depending on the mineral to be extracted, the processes will be different and thus cannot always be generalised.

In a mining project up to four operations can be distinguished (Fig. 3.7), the first of which consisting of exploration and research. This operation seeks to locate mineral deposits in large regions to delimit those areas most likely to contain deposits of interest, depending on the terrain's features. It also seeks to measure the mineral deposit, carrying out detailed investigations, at first on the surface and, then in depth (should the superficial investigations be positive). During the development phase, the necessary resources for mineral extraction, mine design, need for infrastructure, location of the treatment plant, etc. are defined. Next is the most important stage

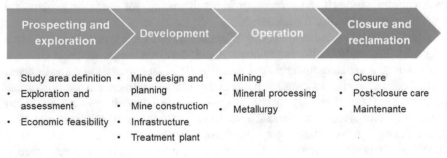

Fig. 3.7 Summary of the life cycle of a mine

from the point of view of the life of the mine: operation, in which the extraction and processing of the ore take place. Finally, we must not forget the closure and reclamation phase, which is aimed at returning the land to the conditions prior to the existence of mining activity.

These phases are usually carried out successively and it should be remembered that, in many cases, the time that elapses between phases is very long, meaning that decades may pass between the exploration phase and the exploitation phase of a mining project. It is worth noting that only around 1% of exploration projects end up becoming mines, either because no mineral deposits have been found or, despite having been found, their extraction is not economically feasible or not possible due to social or environmental reasons. We must bear in mind that the exploration and exploitation of a mining project is a very expensive process, sometimes exceeding hundreds of millions of dollars in the case of large projects, so feasibility studies are crucial.

3.4.1   Exploration and Research

In the past, the exploration and exploitation of mineral deposits focused on those elements that could be easily located on the ground, even with the naked eye. Today, except in some remote and poorly studied regions, this method no longer works. Mineral deposits are scarce and are normally hidden below the surface, and those whose concentration is so high that it makes their extraction profitable are even scarcer. Given that it is impossible to explore in great detail the entire surface of a country or region, in addition to being extremely costly both in terms of money and time, this type of analysis tends to be carried out by mining companies.

Mining exploration and research consist of a series of operations that are geological, geophysical, geochemical, etc. in nature, aimed at discovering and delimiting a mineral deposit to analyse whether it may be feasible to exploit it. In order to minimise the economic risk, exploration is usually carried out in successive stages; in each of them, the areas studied will be smaller and smaller so that, depending on the information obtained, the completion of a phase may or may not lead to the initiation of the next. The terrain can be explored directly, through the study of mineral and rock samples, outcrops and surveys, or indirectly, using, for example, geophysical techniques to study subsoil parameters.

Although there are exceptions, new mineral deposits are sometimes found when originally looking for a different mineral. A well-known example of this is Olympic Dam, a large Australian underground mine. In 1975, an exploration campaign was carried out in the area looking for mineral deposits in sediments in which nothing interesting was found, but, by chance, this great mineral deposit was discovered just below the explored area. This deposit has a very high iron content, of the order of 26%, although not high enough to make it profitable to extract. However, it does contain economically profitable amounts of copper, uranium, gold and silver (Neukirchen & Ries, 2020).

At the end of this first phase, which can last from 1 to 10 years, in the best of cases, the existence of a deposit capable of being extracted will have been revealed. The main characteristics, dimensions, quantity of extractable mineral will be known, semi-industrial metallurgical treatment tests, an economic feasibility study, etc. will have been carried out, and if all the results are positive and if the mining company has the necessary permits and sufficient financing, the next phase of the process can begin.

3.4.2 Development

During the development stage, and after the economic feasibility study, it is possible to continue with the design of the mine, creation of the mine, and with the exploitation phase, where extraction will proceed, leading to the recovery of the mineral from the rock that contains it.

Very few elements in nature appear in an elemental state so that they can be recovered quickly, such as gold or mercury (Fig. 3.8). As a general rule, all metals have to be extracted from mixtures of different minerals that contain them and that appear together in the rock. In the same mine, it is usually possible to extract various elements through different routes and processes that we will explain below. Even in metallic mines, it may be possible to extract non-metallic elements too in the stages prior to the metallurgical process, such as quartz, calcite, barite or feldspar.

Fig. 3.8 Photograph of drops of native mercury in cinnabar, Almadén (Castilla La Mancha, Spain). Author: Miguel Calvo

The size of the mines can vary greatly, from very small (extractions of a hundred tons of ore per day) to large (thousands of tons extracted per day). Furthermore, the opening of a mine involves large investments in machinery and infrastructure that result in considerable social and environmental impacts. All the necessary environmental procedures must therefore be carried out to attempt to mitigate the environmental impacts that will be discussed at the end of this chapter.

Mines can be classified into two types: open-pit mines if the work is carried out on the surface or underground mines if the work is carried out below the surface. Depending on the type of mineral to be mined, its grade, the depth at which the deposit is located, the type of material above the deposit, etc., a particular method will be chosen, or a combination of methods. For example, the ore grade (the concentration of the element of interest, expressed in percentage by weight) can vary greatly from one element to another, from almost 50% in the case of iron to 0.0001% in the case of gold (Neukirchen & Ries, 2020).

In both open-pit and underground mining, mining operations are classified as follows: extraction, ore handling and processing. The most widely used method of extraction, and the most economical, is open-pit mining. In the United States, 97% of metal mining is carried out through this type of mining, and globally this figure is 75%.

In open-pit mines, such as the Chuquicamata mine in Chile, the mineral is exploited by means of horizontal layers called benches, going deeper and wider with each subsequent cut. The thickness of the benches will depend on the type of the deposit, the mineral and the equipment used. This method is used both to extract coal, as well as metallic and industrial minerals. With this type of method, less energy is required per ton of ore, but its impact on the environment is much more significant.

In underground mines, the material is generally extracted through a system of galleries and shafts. It is removed with machinery or with blasting, it is loaded into trolleys or belts and it is taken outside and to the treatment plant. These mines do not greatly alter the landscape, but they are more dangerous due to the risks of collapse and explosions in limited cases. Underground coal mines (where large-scale automation can occur) are also quite common (Fig. 3.9). Also common are mines containing high-value elements, such as gold, whose concentrations mean this extraction process is profitable, despite its high cost.

3.4.3 Exploitation: Extraction and Beneficiation

Of all the processes applied in the mining industry, we can distinguish those that are of a physical type and those that are of a chemical type, all with the final objective of concentrating a specific element or substance. In addition, the physical processes can be further broken down as follows: extraction and storage, crushing, grinding, classification and concentration.

Fig. 3.9 Photograph of one of the coal extraction fronts of the Twentymile mine (USA). Longwall coal cutter that gradually tears coal from the gallery wall; the material falls on a conveyor belt responsible for moving the material to the outside. Author: Peabody Energy, Inc. CC BY-SA 4.0. Wikimedia Commons

The extraction is carried out in the mine, releasing the rock from the rock mass with the goal of recovering the maximum amount of mineral. This process can be achieved using mechanical starting machinery, explosives, etc. As an example, Fig. 3.9 shows how the coal layers are extracted in an underground mine in the United States. Once the material has been extracted, it can be taken to stockpiles where it will rest to reduce the amount of dust generated before taking it to the next treatment stage. This collection also allows the rest of the process to be carried out during periods when no extraction occurs.

The next stage is the mineral processing, which separates the ore (the substance whose extraction is economically profitable), from the gangue or waste rock (the substance whose extraction is unprofitable). These definitions are variable since, as stated before, waste rock can become ore if a change in conditions occurs, for example, if commodity prices increase. The objective of this process is to reduce the amount of material taken to the treatment plant.

Crushing is carried out through a dry process to reduce the particle size of the mineral, with the aim of separating the gangue from the ore, generating a homogeneous material without transforming it into powder. In general, the particle size obtained is around a few centimetres thick. Grinding is carried out through a wet process and seeks to further decrease the grain size of the mineral so that it is more reactive in successive stages of treatment. Using ball mills, the particle size is reduced to a few millimetres. The material with the appropriate particle size is separated through the classification stage, while the rejected material is ground again.

Finally, the concentration process can occur. Separation processes are employed based on the physical properties (electrical, magnetic, gravimetric, etc.) of the elements. An example of this type of concentration process is flotation. In an aqueous

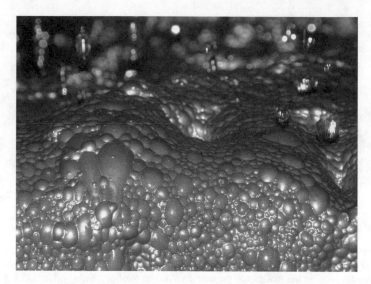

Fig. 3.10 Foam generated at the top of flotation cells during copper extraction at the Prominent Hill (Australia) plant. Author: Geomartin. CC BY-SA 4.0. Wikimedia Commons

medium, bubbles are generated by injecting pressurised air; hydrophobic particles, those that are repelled by water, adhere to these bubbles while hydrophilic ones, those that have an affinity for water, remain in the pulp (Fig. 3.10). In this way, the pulp is concentrated in successive tanks until it reaches the maximum possible recovery. The resulting final pulp can then be passed through thickeners and filters to recover the desired elements.

The energy and water consumption of the stages mentioned above is very high and will largely depend on the grade of the mineral, the hardness of the rock, the ease of separating the mineral from the rock, whether the mine is open-pit or underground, etc. Therefore, it will change considerably from mine to mine and all these factors will influence the total energy consumption. A large portion of energy consumption in mining is associated with milling operations, which are responsible for up to 40% of total consumption (DOE, 2007).

The metallic minerals are also subjected to a smelting and refining treatment. These processes are, in general, very intensive in the use of energy. In fact, their costs predominate over the rest in the industrial chain (from the cradle to the market). There are two basic processes within metallurgy: pyrometallurgy and hydrometallurgy, or sometimes, as is the case with nickel, a combination of both.

3.4.4 Operation: Smelting and Refining

Pyrometallurgy is also known as the smelting and refining process and is based on obtaining and purifying metals through the use of heat, through a dry process. It is a relatively quick process, allowing large quantities of material to be treated in a short amount of time. However, it is not very selective and sometimes it is necessary to repeat it to achieve maximum results. It is also a highly polluting process and consumes a large amount of energy.

The Encyclopaedia Britannica defines smelting as "the process by which a metal is obtained, either as the element or as a simple compound, from its ore, by heating beyond the melting point, ordinarily in the presence of oxidising agents, like air, or reducing agents, such as coke". The refining process typically refers to electrorefining, a method by which a metal is purified by electrolysis.

Before smelting, the extracted rock must undergo crushing, grinding and a concentration process. Furthermore, in the case of carbonates and sulphides, these require a prior roasting or calcination process, which converts them into the corresponding metal oxide. For example, this is the case of malachite (copper carbonate), limestone (calcium carbonate) and magnesite (magnesium carbonate), which undergo a simple thermal decomposition, releasing CO_2 into the atmosphere and leaving the metal oxide free. In other cases, such as that of galena, it is first converted to sulphate and then thermally decomposed into lead oxide and gases that decompose to form sulphuric acid.

Each smelting process varies depending on the mineral's nature, the metal involved and the purity of the final product. However, all require a high-temperature furnace to which it is necessary to add reducing agents. The furnace produces impure molten bullion and waste materials, the so-called slags, which are easily separated by density difference. In the case of sulphides, it is also possible to obtain economically important by-products such as gold and silver.

The last stage of pyrometallurgy is refining, which consists of separating the main metal from other elements present in the bullion that has been generated in the furnace. For this, methods such as electrorefining can be used: an electroplating process, by which the unpurified metal constitutes the anode, which is dissolved in an acidic electrolyte transporting medium. When an electric current passes through this medium, the dissolved metal ions migrate to the cathode (Fig. 3.11), where they deposit as a pure metal.

It is very important to understand the behaviour of the elements in order to recover valuable metals such as nickel, copper and cobalt, and also to be able to eliminate harmful impurities such as arsenic, antimony and bismuth, since the processes described above take place at very high temperatures and release volatile and highly toxic gases, including those of heavy metals (Tan & Neuschutz, 2001).

Fig. 3.11 Stainless steel cathodes where the copper will be deposited during the refining process. Author: Xstrata Technology. CC BY-SA 4.0. Wikimedia Commons

3.4.5   Exploitation: Refined Hydrometallurgy

Hydrometallurgy is based on the obtaining and purification of metals through a wet process, by means of reactions in an aqueous phase at low temperature. It is a process that generates less environmental impact from an energy point of view and has a lower investment cost than pyrometallurgy. In addition, the plants are modular in nature, facilitating expansion from a small-scale operation to a medium-sized operation. There is great automation, and there is great control over the chemical reactions that occur due to the kinetic conditions in which these processes take place.

Hydrometallurgy can be used for the recovery of minerals, concentrates, recycled or residual materials. It offers a good alternative in terms of energy savings by taking advantage of aqueous chemistry by leaching metals. It is a more selective process than pyrometallurgy since reactions can be generated in which only part of the ore dissolves. It is divided into several successive stages which are leaching, concentration and purification and precipitation.

Leaching is the process that includes the chemical attack, in an aqueous phase, of the metal to be recovered. This attack can be acidic, basic or neutral, depending on the mineral, although sulphuric acid is used in many processes. Leaching can be done in situ, in accumulated ore piles or ponds (Fig. 3.12), collecting the generated solution, or in dedicated tanks.

After leaching, the next stage is purification and/or concentration, carried out on the solution obtained in the previous stage, which seeks to eliminate certain

Fig. 3.12 Leaching ponds at the Mopani Copper Mines copper mine (Zambia). Author: Photosmith 2011. CC BY-SA 2.0. Wikimedia Commons

impurities from the solution prior to the precipitation stage. Chemical methods which favour the precipitation of these impurities, cementation, extraction with solvents, etc. are usually employed. The last stage, that of precipitation, seeks to separate the valuable metal, in its elemental or oxidised form, from the solution. An electrolysis or electrodeposition process is usually carried out, where cathodes of the pure metal are generated, as explained above.

Bio-hydrometallurgy is an intermediate between biotechnology and hydrometallurgy, which uses microbes for bioleaching and bio-oxidation of metal minerals (Rossi, 1990). In theory, almost any chemical process can be carried out with bacteria and this opens up a wide field for research and development. Such processes are now used, for example, for the treatment of heavy metals and wastewater, as well as for the recovery of high-value metals such as copper, nickel, lead, zinc and gold. In fact, today over 15% of total copper production in the United States comes from leaching with bacteria. Much of this copper comes from leaching in mine tailings, but more and more research is being undertaken on in situ leaching with bacteria, in which bacteria-laden waters are pumped over the mineral deposits, recovering the generated liquid.

Bacterial mineral leaching processes require little energy and offer many benefits, such as accessibility, in remote locations. However, for most of the mining processing industry, this technology has limited commercial application since it is a slow process and the limits of these bacteria have yet to be determined (Sarveswara Rao & Acharya, 2008).

Leaving bacterial leaching aside, hydrometallurgy is gaining ground on pyrometallurgy due to its lower environmental impact. Some authors claim that hydrometallurgical processing has proven to be highly innovative and economical (Lakshmanan et al., 2003). However, hydrometallurgy, like pyrometallurgy, also has a few disadvantages, the most significant being water contamination. This process generates muds with unrecovered materials which end up in tailing ponds. In some extreme cases, this can lead to dam failures and contamination of the adjacent lands, contamination of groundwater, etc. An example of a waste rock pond that unfortunately became well known is the one that took place in 1998 at the Aznalcóllar mine, Seville (Spain), considered one of the largest environmental disasters in Europe. This dam burst caused acidic waters and toxic sludge to flood an area of almost 62 km in length, also affecting the Agrio and Guadiamar rivers, numerous aquifers and even the Doñana National Park (Ayala-Carcedo, 2004).

3.4.6 Case Study: Copper Processing

Having seen the general processes, we will now delve into the process of an important metal: copper. Both pyrometallurgy and hydrometallurgy can be used, although only 10% of the primary copper comes from the latter process.

The pyrometallurgical process is applicable when dealing with mixtures of primary and secondary copper sulphides such as chalcopyrite, bornite and enargite, among others. The minerals, once extracted, go through the crushing and grinding stages to reduce their particle size. Subsequently, they undergo flotation to concentrate the copper. This process brings the copper content from 2–5 to 20–45% in the concentrates obtained after flotation. In addition to copper concentrate, minor amounts of other elements such as antimony, bismuth, selenium, cobalt, tellurium, gold, palladium, etc. may appear in this pulp (BCS, 2002).

After pressing and drying, the resulting product undergoes smelting to extract the copper in a series of successive stages (Fig. 3.13). The roasting process, separated

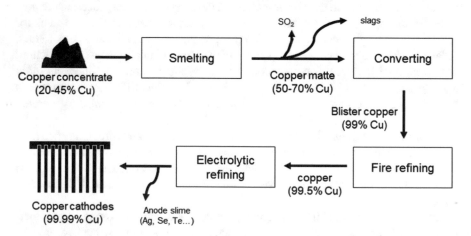

Fig. 3.13 General pyrometallurgical process of copper. Figure prepared from IPPC data (2017)

or integrated into the smelting process, heats the copper concentrate and partially oxidises the chalcopyrite, also reducing impurities. The most common smelting process is completed in the oxygen flash furnace (Mullinger & Jenkins, 2008). This furnace, which runs on natural gas and air at 1,000–1,100 °C, combines roasting with smelting and converts chalcopyrite into copper sulphide and ferrous oxide. The raw materials, in addition to chalcopyrite, are limestone and sand. The excess oxygen in air selectively reacts with iron to produce iron oxide while liberating copper sulphide (II). Later, the excess sulphur reduces copper(II) to copper(I) sulphide which melts. The silicon dioxide in sand reacts with the limestone and the iron oxide to produce a slag predominantly made of ferrous and calcium silicates. This slag floats in the molten copper sulphide, referred to as matte, which has a copper content of between 50 and 70%. Slag and matte can be recovered separately at the bottom of the furnace. Copper oxide can also be recovered in the process, from the slag leaving the flash furnace.

The copper matte is then treated in a continuous converter furnace with oxygen-rich air. The copper sulphide and the remaining iron sulphide are oxidised to copper, iron oxide and SO_2 in a highly exothermic reaction. This results in a material called blister copper that has a 99% copper purity, in addition to slag.

The blister is then introduced into refining furnaces, where the rest of the sulphur and other impurities that may have remained are removed, and anodic copper with a purity of 99.5% is obtained as a result.

This copper anode undergoes electrolysis in cells containing an acidic copper sulphate solution for seven days so that the copper migrates from the anode to the cathode, depositing there and obtaining copper cathodes with a purity of 99.99%. Anodic slimes are where some precious metals such as silver, selenium, tellurium, etc. are concentrated, which can later also be recovered through other treatments (Lossin, 2000; Moskalyk & Alfantazi, 2003; IPPC, 2017). For every ton of copper ore that is mined, 1.5 tons of slag and 2 tons of sulphur dioxide are produced.

Copper can also be produced from secondary sources. However, the other materials and metals that accompany it, such as plastic coatings and oils, must also be taken into consideration. Several pre-treatments, such as reducing organic volatile compounds, minimising dioxins and furnace emissions must be also executed. Secondary smelting of copper uses coke to maintain reducing conditions in the furnace since the raw material is either metallic or oxidised. The furnaces volatilise the zinc, tin or lead from copper alloy scrap. Such metals are then collected in the filter dust and the off-gases containing SO_2 and CO_2 are wet-scrubbed to produce sulphuric acid.

A minimum percentage of the copper mineral is produced by hydrometallurgy, especially in the case of copper oxides, such as malachite, azurite, chrysocolla or cuprite, among others. After crushing and grinding the mineral, leaching is carried out using sulphuric acid and water (Fig. 3.14).

This process is usually carried out through heap leaching, which consists of stacking large quantities of mineral in piles that are then watered with diluted solutions of sulphuric acid. A copper sulphate solution is thus generated and gradually recovered, with copper concentrations of up to 9 g per litre. This process can take

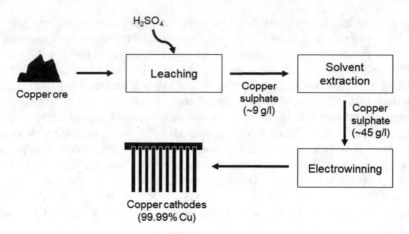

Fig. 3.14 General hydrometallurgical process for copper. Figure prepared from IPPC data (2017)

months or even years, until the copper in the material pile runs out. After leaching comes purification, via extraction with solvents and the adding of organic chemical reagents that separate it from the rest of the solution. The resulting organic solution is subjected to re-extraction in order for the metal to re-enter the aqueous phase. This results in a copper sulphate solution with up to 45 g per litre of copper. Finally, the solution is taken to large plants with stainless steel cathodes and inert anodes in the electrolysis and electrowinning stage, obtaining (approximately one week later) solid copper deposited on the cathode with a purity of 99.99%.

3.4.7 The Wheel of Metals

Most of the metals that are mined today cannot be recovered independently of each other, neither hydrometallurgically nor pyrometallurgically. The production or recovery of a metal is usually linked to or dependent on that of another. A well-known example discussed above being that of gallium, an element that is extracted as a by-product of bauxite processing and zinc processing residues. And it is not the only one: many other minerals are mined as by-products from base metal mining. For example, PGM usually appears associated with nickel sulphides, and indium is usually obtained from refining zinc minerals. In other cases, the elements may substitute the main metals, such as zinc ore, sphalerite (where indium can substitute zinc), or in the case of molybdenum ore, molybdenite (where rhenium can substitute molybdenum).

In order for these secondary metals, which appear in very small quantities, to be recovered, it is often necessary for them to undergo one or more concentration processes before their extraction is profitable. This concept is very well illustrated in the wheel of metals that appears in Fig. 3.15 (Verhoef et al., 2004). In it, the elements

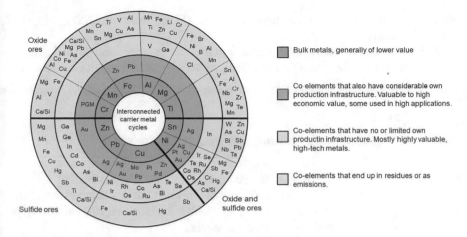

Fig. 3.15 Wheel of metals showing the links between the different elements (Verhoef et al., 2004)

that appear in the central circle, such as copper (Cu), aluminium (Al), manganese (Mn) or nickel (Ni), are the main metals that are extracted and also those that are consumed in greater quantity today.

The following three circles represent those minor elements that usually appear associated with each of them. For example, if we start with tin (Sn), silver (Ag) appears in the next circle (green) and indium (In) appears in the subsequent circle (yellow), all frequent by-products in this type of mining. If we continue with the processing and refining, from tin, minerals bismuth (Bi), zinc (Zn), copper (Cu), and many other metallic elements could also be recovered according to the outer circle.

Well-designed metallurgical processes of each segment can cope with the optimal economic, technological and environmental recovery of the corresponding minor elements. However, on many occasions, the elements of the outermost circle are lost, cannot be recovered, and end up as waste. This happens, for example, in the case of lead and cobalt, which appears with manganese oxides; iron is also lost in copper metallurgy, and vice versa.

Taking the example of copper, in anodic slimes there can be up to 10% selenium (Se) and 5% tellurium (Te), in addition to other precious metals. Concerning these two mentioned elements, they are recovered by roasting slime pellets with soda ash at around 550–650 °C. This process oxidises both elements to their hexavalent state and the resulting product is then leached in water. Sodium selenate dissolves, but sodium tellurate, which is insoluble in alkaline solution, can be filtered out. The dissolved sodium selenate is then reduced to form sodium selenide, which is leached further to produce a selenium slurry. The resulting precipitate is melted and passed through a sieve, recovering the metallic selenium (Ally et al., 2001). In the case of tellurium, it can be recovered from copper anodic slimes by cementation with copper, forming copper telluride which is then leached with sodium hydroxide and air to produce a sodium telluride solution. A controlled oxidative process follows which can produce either tellurium in metal form or tellurium oxide (Hoffmann et al., 2000; Langner 2000; Fthenakis et al., 2007).

Therefore, both the elements that appear in the main minerals and those that may be contained in smaller quantities and that can only be used through very specific refining stages influence the final recovery.

3.4.8 Reclamation, Rehabilitation and Post-closure

Reclamation of a mine once it is exhausted is a universally accepted standard, at least in developed countries. The exact details of how this reclamation should be carried out depend on the legislation in force in each country, international ISO standards or even the policies of the different mining companies.

Environmental reclamation must be carried out by the mining company with its own means, and the associated costs must also be covered by the company. In some cases, these costs must be set aside in advance to guarantee the environmental recovery of the affected land if the company closes early or goes bankrupt. Similarly, the operating entity is obliged to take all necessary measures, based on the best available techniques, to prevent or reduce as far as possible any negative effect on the environment and on human health.

Furthermore, reclamation activities should include backfilling the holes left by open-pit mining, carrying out adequate topographic modelling, avoiding large slopes and steep slopes, recovering the topsoil and revegetating, with local flora, restoring and improving watersheds, avoiding drainage, etc. They should also include a series of measures to rehabilitate the services and facilities in the vicinity of the mines: treatment plants, offices and other fixed buildings, for instance.

In some cases, when an open-pit mine cannot be restored to the initial conditions of the side due to its great depth or another characteristic, other alternatives can be explored. An example of this is the old lignite mine in Cerceda, in Galicia (Spain), which was filled to create the world's largest artificial lake. A similar case is that of the As Pontes lignite mine, also in Galicia, a lake that is currently used for recreational use.

3.5   Environmental Impacts of Mining

Mining is one of the activities with the greatest environmental impacts if the complete cycle from cradle to grave is considered. Open-pit mining currently predominates, and as mineral concentrations in the mines decrease, the amounts of material that must be extracted to obtain the same amount of metal increase. Energy, water and reagents consumption also increases.

Mining modifies the landscape and has long-term impacts on both communities and natural resources, disturbing the land, watersheds, etc. The disposal of waste rock, tailings and toxic materials, acid mine drainage or energy consumption and the associated CO_2 emissions are major health concerns for both mine workers and those living in the vicinity. Furthermore, the effects on the environment are notably higher in the case of open-pit mining. This is due to a relative increase in the amount of material to be moved, the use of heavy machinery and the size of this machinery. As the demand for minerals grows and the mines with higher concentrations are depleted, a greater quantity of material will have to be removed and the effect on the environment will intensify.

In a mining operation, both open-pit and underground, the dynamics of local watersheds can be modified. In fact, water is the first resource that is affected by mining activity, and it also serves as a vehicle for contamination.

Water quality can be affected by mine discharges, such as drainage water and wastewater from mining facilities. The lower the grade in the mine, the more water needs to be used. For mining companies, the costs of owning large water treatment plants can make the mine unprofitable, as the relationship between decreased mine grade and water demand is not linear but exponential. Therefore, the demand for water and its treatment can be a determining factor in the closure of a mine. The water involved in the mining process may contain metals in low concentrations and toxic residues from the separation processes. In some cases, even very low concentrations of certain elements can be extremely damaging to aquatic ecosystems and agriculture, which only exacerbates social and environmental problems.

The location of the exploitation hole, in the case of open-pit mines, as well as auxiliary infrastructures (sludge ponds, dumps, tailings, offices, workshops, toilets, treatment plants, etc.) that occupy large extensions of land, can directly affect aquifers, wetlands and surrounding rivers with the consequent need to divert them.

Finally, we must not forget the reduction in the availability of water as a resource. Groundwater pumps will be used to facilitate the extraction of the mineral in the case of underground mining. This can significantly affect the flow of the river channels of nearby towns' environment and supply systems.

In the case of metallic mining, which includes obtaining all those minerals whose objective is the subsequent extraction of metals (iron, copper, tin, zinc, lead, gold, silver, etc.), the mineral, once extracted, must be treated in plants for its processing and separation. In general, as we have seen, this process involves crushing and grinding to decrease the grain size of the rock and release the mineral, and a concentration phase where physical and chemical processes are used. Processing generates wastes that, if accumulated and abandoned, can produce significant negative effects on the environment, including water acidification. This process occurs when there are large amounts of sulphides in the mines, which oxidise in the presence of oxygen in the air and water, generating leachates that considerably decrease the pH of waters. These waters can accumulate in abandoned mine holes, generating large ponds that are both highly acidic and toxic, or circulate freely until they reach rivers or groundwater. The environmental consequences of acidification of the surrounding waters will depend on their pH, which may cause malnutrition of the soils (due to leaching phenomena). This in turn hinders vegetation growth and the development of habitats that serve as shelter and food for native fauna.

The acidification phenomena of the waters usually also entail the mobilisation of the metals contained in the different layers of the soil, with the possibility that the heavy metals are dispersed in non-mineralised areas. The presence of these heavy metals in the waters can even affect aquifers in the environment or water supply systems.

A striking case is that of Andalusia (Spain), specifically in the Iberian Pyrite Belt. Here, the presence of metallic minerals has caused the presence of very acidic, metal-filled waters in the area for centuries, to the point of the rivers being given names reflective of the colour of their waters, a well-known example being Río Tinto (Fig. 3.16). Human activity has only intensified this process, as more mineral matter is exposed to the air. In addition, the presence of abandoned unrestored mines exacerbates this impact, particularly in the landscape and in the waters.

Regarding the atmosphere, the effects of mining activities on the environment are mainly focused on three fundamental aspects: the production of dust, noise (and associated vibrations) and the generation of atmospheric pollutants.

Fig. 3.16 Photograph of Río Tinto: the reddish colour of its waters that gives it its name is due to the acid drainage of the sulphide-rich soils in the area. Author: Antonio. CC BY-SA 2.0. Wikimedia Commons

Fig. 3.17 Large truck used to transport material from the mining area of Cripple Creek (United States). Author: James St. John. CC By 2.0. Wikimedia Commons

In the case of dust, it mainly comes from road traffic on tracks and unpaved areas (Fig. 3.17), from the loading and unloading of ore, from explosions to remove material and from the erosive action of the wind on dumps and stockpiles. This dust is especially undesirable when it is generated in areas close to urban centres or communication routes. Similarly, this transport, blasting and other mine processes can generate noise and vibrations.

The consumption of fossil fuels, either for the generation of electricity or for the transport of the material, emits enormous amounts of greenhouse gases, such as carbon dioxide, nitrogen oxides and particles in suspension. In the future, and within the framework of sustainable mining, mining facilities will have to apply carbon capture and storage or use alternative means of transport and cleaner fuels. Mining is responsible for between 8 and 10% of global energy consumption; of this, around 17% is used for transport activities and a significant portion is used by ventilation and pumping in the case of underground mining (DOE, 2007). The energy consumed in a mine largely depends on the concentration of the mineral, increasing considerably when mineral concentration decreases.

In addition to emissions from the consumption of fossil fuels, mining companies can emit other much more dangerous substances into the atmosphere, such as mercury or lead. Lead can cause neuropsychological damage and cognitive defects, especially in young children, while mercury, if inhaled, can be a very powerful neurotoxin. According to the United States Toxic Release Inventory, the metal mining industry is responsible for 9% of all industrial mercury emissions in the country (Toxic Release Inventory, 2019). In Spain, the Almadén mining district in Ciudad Real was one of the main suppliers of mercury worldwide from the Roman era to 2002, when the mines closed due to the drop in price and the reduction in the use of mercury due to its high toxicity. In addition to the natural contamination generated by the presence of this element in soils and rocks, human action and especially mining activity have exacerbated dispersal, generating soils and waters with mercury concentrations that in some cases, exceed the recommended amounts by a factor of several hundred.

In the case of underground mining, the effects on the atmosphere are considerably lower from an environmental point of view since most of the activities take place underground. Nonetheless, infrastructures continue to exist abroad: stockpiles of land, machinery traffic, etc., which can produce dust, noise and greenhouse gas emissions.

In the extractive industry, soil conditions are easily identifiable. In an open-pit mining operation, the mineral-free layer of soil must be first removed in order to extract the mineral. Significant amounts of material must therefore be removed, ending up in dumps and causing the soil to lose its original topographic profile, its composition and function when the surface layers are removed or covered by other inert materials. Meanwhile, the soil is increasingly eroded due to greater fragmentation, leading to surface runoff and, ultimately, affecting water quality. This causes an increase in the level of turbidity or the amount of solids in suspension the water carries.

This inevitably affects the natural landscape due to the drastic changes it is subjected to, particularly in the case of open-pit mining. The creation of vast mining holes, tailings and ponds are determining causes of environmental degradation. In the case of underground mining, this effect on the landscape is reduced, although it still has considerable impact, particularly in terms of subsidence. By creating a vacuum in the subsoil caused by mineral extraction, the land's balance is altered. Complex movements of an entire area can occur, triggering both superficial and deep subsidence.

The consequences on local fauna and flora are also difficult to avoid as land use is modified. Furthermore, the presence of machinery, noise emissions and vibrations contribute to the negative effects of mining on the surrounding environment.

In terms of mining activity, the effects caused by the metal beneficiation process are also considerable. In the smelting process a reducing agent—a solvent that facilitates the reaction—is typically employed to melt the mineral, in addition to high temperatures. The required temperature is obtained from the burning of fossil fuels, normally coal. Coal is typically used as the reducing agent and limestone as the solvent. Electricity is also needed if electrolysis is carried out, typically generated by consuming more coal. This process results in large amounts of CO_2 emissions, in addition to other pollutants emitted into the air, such as sulphur dioxide from sulphide smelting or hydrogen fluoride from aluminium processing. These gases can also generate smog, ozone, oxides of nitrogen and sulphur, carbon monoxide and particles in suspension, contaminating the surrounding air, soil and waterways. Also notable is the large amount of sulphuric acid used in the production of many metals, leading to acid rain.

The slags formed during the smelting process can contain non-negligible amounts of heavy metals such as cadmium, lead, arsenic, nickel, copper or zinc, which require adequate treatment to prevent these particles from ending up in the air or water. These slags are increasingly being used to obtain these elements, through a post-treatment process, and may also present a recycling opportunity in the future.

Strict environmental regulation in many countries has led many companies to use more efficient technologies and has led the new smelting plants greatly to lower emissions. Electrostatic precipitators and desulphurisation plants have reduced and even eliminated some environmental impact, especially in relation to heavy metal-laden dust and acid rain. They have also solved the problem of bioaccumulation associated with particles emitted in foundries, an issue widely researched in the 1980s and 1990s (Reichrtova et al., 1989; Leita et al., 1991).

We should remember that, despite the great impact of mining, there has been an increasing tendency in recent years to attempt to mitigate some of this impact through the use of renewable energy sources and energy efficiency improvements in the processing stages. Specifically, in Latin America, there are many countries and mining companies concerned with finding alternative sources to fossil fuels using hydroelectric energy and solar plants. An example of this is the Pampa Elvira solar plant, owned by the CODELCO mining company, which annually generates 51,800 MWh of heat used during the process of electro-obtaining copper cathodes from the Gabriela Mistral Division (Nelson, 2017). There are also other more sustainable alternatives, such as the use of biofuels in mining or biomass trains to produce electricity, as happens in Minera Tres Valles, a mining project in Chile (Van Den Hurk, 2017).

3.6 Social Impacts of Mining

The environment is not the only area affected by mining and beneficiation; social impact must also be considered. In addition to the possible effects that emissions and soil and water contamination can have on the health of people living near these mines, the effects of related issues such as socio-environmental conflicts, population movements and illegal mining must also be assessed.

In Latin America, a region with great diversity and mineral wealth, there are many social conflicts generated due to mining. There is even a platform called the Observatory of Mining Conflicts of Latin America (OCMAL[1]), where interested parties can access a map of the area where the different conflicts are located geographically. Colombia occupies one of the leading positions in the ranking of countries with the highest rates of forced displacement worldwide. This is a situation that the mining industry has promoted, with the most affected areas being those belonging to peasants, indigenous people and Afro-descendants (Betancur Vargas & Pérez Osorno, 2016). Illegal mining is a widespread issue across Latin America, especially in the case of gold extraction, a high-value metal. Specifically, between 80 and 90% of the gold produced in Venezuela comes from illegal mining. This figure is 80%

---

[1] Observatory of Mining Conflicts of Latin America: https://www.ocmal.org.

in Colombia, 77% in Ecuador and 30% in Bolivia (The Global Initiative Against Transnational Organized Crime, 2016). This illegal gold mining has serious negative effects, displacing not only the local population but also contaminating the waters by using mercury in the extraction process. Mercury is burned via uncontrolled and rudimentary techniques, destroying ecosystems and areas protected by deforestation and modifying the soil profile.

At a global level, other initiatives exist that measure and evaluate the conflicts generated by the mining sector, particularly the Atlas of Environmental Justice,[2] launched in 2014 as the result of a project financed by the European Commission (Temper et al., 2014). In this atlas we can see the conflicts categorised by raw materials and type of conflict (water use, biodiversity, access to the territory, waste management, etc.). This shows us that a wide variety of environmental, social and economic conflicts exist, as well as recurring effects regardless of where the mine is located. These recurring effects match a series of patterns that can be extrapolated to any country, whether developing or otherwise.

Mining is necessary to meet the global demand for minerals and metals, although as we have seen it entails a series of associated effects which vary depending on the regions and the restrictions set by the local legislation. Empirically speaking, new mineral deposits with high element concentrations and whose extraction is economically feasible are increasingly unlikely to be discovered. Technological advances can make it possible to extract a mineral with lower concentrations, but the energy required—and the associated environmental and probably social impacts—will be much greater.

A model like *Thanatia*, which evaluates the depletion of resources from cradle to grave and back to cradle, taking into account its scarcity in the crust and the energy consumption of the entire chain, will allow us to analyse criticality from a new perspective. It allows us to not only focus on what type of minerals are needed in the present or will be in the future but also those that will be scarcer or more difficult to recover until their initial state. The question that arises then is how we can assess this loss of mineral wealth—something that can be tackled through thermodynamics.

References

Ally, M. R., Berry, J. B., Dole, L. R., Ferrada, J. J. & Van Dyke, J. W. (2001). *Economical recovery of by-products in the mining industry*. Available at: http://www.lesdole.com/TM225.pdf.
Ayala-Carcedo, F. J. (2004). La rotura de la balsa de residuos mineros de Aznalcóllar (España) de 1998 y el desastre ecológico consecuente del río Guadiamar: Causas, efectos y lecciones. *Boletín Geológico y Minero, 115*(4), 711–738.
BCS. (2002). *Energy and environmental profile of the U.S. mining industry. Mining industry of the future, Copper*. Available at: https://www.energy.gov/sites/default/files/2013/11/f4/copper.pdf.

---

[2]Atlas of Environmental Justice: https://ejatlas.org.

Betancur Vargas, A., & Pérez Osorno, M. M. (2016). Desplazados legales o ilegales: Una mirada desde los procesos extractivos en Colombia y contexto general de algunos países latinoamericanos. *Anuario Latinoamericano Ciencias Políticas y Relaciones Internacionales, 3*, 241–273.

BP. (2018). *Energy outlook* (2018 ed.). Available at: https://www.bp.com/content/dam/bp/business-sites/en/global/corporate/pdfs/energy-economics/energy-outlook/bp-energy-outlook-2018.pdf.

BP. (2019). *British petroleum statistical review of world energy*. Available at: https://www.bp.com/content/dam/bp/business-sites/en/global/corporate/pdfs/energy-economics/statistical-review/bp-stats-review-2019-full-report.pdf.

Bustillo Revuelta, M. (2018). *Mineral resources, from exploration to sustainability assessment*. Springer.

Calaway, L., & van Rensburg, W. C. J. (1982). US strategic minerals: Policy options. *Resources Policy, 8*(2), 97–108.

Calvo, G., Valero, A., & Valero, A. (2017). Assessing maximum production peak and resource availability of non-fuel mineral resources: Analyzing the influence of extractable global resources. *Resources, Conservation and Recycling, 125*.

Campbell, C. J. (1997). Better understanding urged for rapidly depleting reserves. *Oil and Gas Journal, 7*, 51–54.

DOE. (2007). *Mining energy bandwidth analysis process and technology scope*. Available at: https://www.energy.gov/sites/prod/files/2013/11/f4/mining_bandwidth.pdf.

Fthenakis, V.M., Kim, H.C., & Wang, W. (2007). *Life cycle inventory analysis in the production of metals used in photovoltaics*. United States. https://doi.org/10.2172/909957.

Gaswirth, S. B., French, K. L., Pitman, J. K., Marra, K. R., Mercier, T. J., Leathers-Miller, H. M., Schenk, C. J., Tennyson, M. E., Woodall, C. A., Brownfield, M. E., Finn, T. M., & Le, P. A. (2018). *Assessment of undiscovered continuous oil and gas resources in the Wolfcamp Shale and Bone Spring Formation of the Delaware Basin, Permian Basin Province, New Mexico and Texas, 2018*, Reston, VA.

Hoffmann, J. E., Reimers, J., & Schlewitz, J. E. (2000). Kirk-Othmer encyclopedia of chemical technology (pp. 1–27). John Wiley & Sons.

International Energy Agency. (2018). U.S. total crude oil and products imports by country of origin. https://www.eia.gov/dnav/pet/pet_move_impcus_a2_nus_ep00_im0_mbbl_m.htm.

IPCC. (2000). *Special report on emissions scenarios: A special report of working group III of the intergovernmental panel on climate change*. Cambridge University Press.

IPPC. (2017). *Best available techniques (BAT) reference document for the non-ferrous metals industries*. Available at: https://eippcb.jrc.ec.europa.eu/sites/default/files/2020-01/JRC107041_NFM_bref2017.pdf.

Ivanhoe, L. F., & Leckie, G. G. (1993). Global oil, gas fields, sizes tallied, analyzed. *Oil and Gas Journal, 91*(7), 87–91.

Laherrère, J. H. (1994). Published figures and political reserves. *World Oil, 33*.

Lakshmanan, V. M., Sridar, R., Harris, G. B., & Ramachandran, V. (2003). *Innovation: The way forward for hydrometallurgical processing*. In C. Young, A. M. Alfantazi, C. G. Anderson, D. B.Dreisinger, B. Harris & A. James (Eds.), *Hydrometalurgy 2003—Fifth International Conference* (p. 1415).

Langner, B. E. (2000). *"Selenium and selenium compounds," Ullmann's encyclopedia of industrial chemistry*. Wiley-VCH Verlag GmbH & Co. KGaA.

Leita, L., Enne, G., Nobili, M., De, B. M., & Sequi, P. (1991). Heavy metal bioaccumulation in lamb and sheep bred in smelting and mining areas of S.W. Sardinia (Italy). *Bulletin of Environmental Contamination and Toxicology, 46*, 887–893.

Lossin, A. (2000). *"Copper," Ullmann's encyclopedia of industrial chemistry*. Wiley-VCH Verlag GmbH & Co. KGaA.

Moskalyk, R. R., & Alfantazi, A. M. (2003). Review of copper pyrometallurgical practice: Today and tomorrow. *Minerals Engineering, 16*, 893–919.

Mullinger, P., & Jenkins, B. (2008). *Industrial and process furnaces. Principles, design and operation*. Butterworth-Heinemann (Elsevier).

Nelson, I. (2017). La experiencia de minera Codelco. *Energía y Negocios, 112*, 18–20.

Neukirchen, F., & Ries, G. (2020). *The world of mineral deposits*. Springer.

Perello, J., & Cabello, J. (1989). Pórfidos cupríferos ricos en oro: Una revisión. *Revista Geológica De Chile, 16*(1), 73–92.

Reichrtova, E., Oravec, C., Palenikova, O., Horvathova, J., Takac, L., & Bencko, V. (1989). Bioaccumulation of metals from a nickel smelter waste in P and F1 generations of exposed animals. II. Modulation of immune processes. *Journal of Hygiene, Epidemiology, Microbiology, and Immunology, 33*(2), 245–251.

Rossi, G. (1990). *Biohydrometallurgy*. McGraw-Hill.

Sarveswara Rao, K., & Acharya, S. (2008). Trends in non-ferrous hydrometallurgy-solutions and challenges. In C. Young, P. R. Taylor, C. G. Anderson & Y. Choi (Eds.), *Hydrometalurgy 2008—Sarveswara Rao 2008* (pp. 416–424).

Tan, P., & Neuschutz, D. (2001). A thermodynamic model of nickel smelting and direct high-grade nickel matte smelting processes: Part II. Distribution behaviors of Ni, Cu Co, Fe, As, Sb, and Bi. *Metallurgical and Materials Transactions B, 32*, 341–351.

Temper, L., Bene Del, D., Argüelles Cortés, L., & Çetinkaya, Y. (2014). EJAtlas, mapeo colaborativo como herramienta para el monitoreo de la (in)justicia ambiental. *Ecología Política, 48*, 10–13.

The Global Initiative Against Transnational Organized Crime. (2016). *Organized crime and illegally mined gold in Latin America*. Available at: https://globalinitiative.net/analysis/organized-crime-and-illegally-mined-gold-in-latin-america/.

Tornos Arroyo, F. (2008). La geología y metalogenia de la Faja Pirítica Ibérica. *Macla, 10*, 13–23.

Toxic Release Inventory. (2019). *U.S. toxic release inventory*.

USGS. (2019). *Mineral commodity summaries 2019. United States Geological Service*. Available at: https://prd-wret.s3-us-west-2.amazonaws.com/assets/palladium/production/atoms/files/mcs2019_all.pdf.

USGS. (2020). *Mineral commodity summaries 2020. United States Geological Service*. Available at: https://pubs.usgs.gov/periodicals/mcs2020/mcs2020.pdf.

Van Den Hurk, A. (2017). The quadrant of mining and renewables. *Australian Mining, 10*.

Verhoef, E. V., Dijkema, G. P. J., & Reuter, M. A. (2004). Process knowledge, system dynamics, and metal ecology. *Journal of Industrial Ecology, 8*(1–2), 23–43.

Chapter 4
The (Thermodynamic) Value of Scarcity

Abstract For millions of years, nature has formed and concentrated minerals in the deposits we know today. Such deposits represent the natural stock or cradle from a life cycle assessment point of view. The conventional cradle-to-grave analysis can be combined with grave-to-cradle analysis using *Thanatia*, a resource-exhausted planet, as a starting point. Using exergy replacement costs (ERC) through thermodynamics, we can have an idea of the amount of energy that would be necessary, using conventional technologies, to achieve concentrations in mineral deposits, from Thanatia. If we add ERC to the energy associated with the mining process, beneficiation and processing, we can also assess mineral resources' thermodynamic rarity. This concept can be used then as an indicator of scarcity. Accordingly, we can compare it with other criticality indicators, such as those established by the European Commission, or use it to physically assess the quality of raw materials found in technologies properly.

The continued extraction of materials from the Earth's crust implies an irreparable net reduction of the natural stock. Consequently, the Earth gradually loses its mineral wealth, reaching a hypothetical state of a degraded planet with no mines or concentrated minerals. We call this planet Thanatia and it is the basis for measuring the rate of depletion of mineral resources through thermodynamics. We will talk about the two main laws of thermodynamics, which will serve as a basis for us to assess the loss of mineral wealth later.

In this section, we will establish the basis of the methodology used for this purpose, physical geonomy. One of its objectives is to quantify the costs avoided by currently having minerals concentrated in mines and not dispersed as they would be on a degraded planet. To do this, first, we need to define the life cycle of materials.

4.1 The Life Cycle of Materials

The planet functions through solar energy-powered cycles, including those of carbon, oxygen, nitrogen, phosphorous, sulphur and water. There are certain elements, such as some metals, that are associated with biota that have short cycles and that can be quickly replaced, as long as the production of waste from anthropogenic sources does not exceed the assimilation capacity of the biosphere. In nature, many of these cycles are closed: resources (once used by organisms) end up as waste and are re-used by other organisms, and so forth. However, so far there are no defined cycles for minerals. Dispersion is a problem associated with the cycles of minerals directly caused by humans. The waste generated is rarely reused and thus these materials are accumulated in landfills.

In the event that we wanted to restore the original mineral deposits as they were before their extraction, the simplest way to do this would be using fossil fuels. Still, these fossil fuels have also taken millions of years to form through their own cycles, posing the same problem. Therefore, it is important to know the metabolism of minerals in our society. The most widely used method is life cycle assessment, which considers the cycle of materials from their extraction (cradle) to the final product and landfills, known as the grave.

Two international standards, ISO 14040 and ISO 14044, describe the requirements and recommended elements that a life cycle assessment should include identifying four phases: (1) definition of the scope and objectives; (2) life cycle inventory assessment; (3) life cycle impact assessment; and (4) interpretation.

The first phase, defining scope and objectives, serves to determine the depth at which the study will be carried out. The study's scope will serve to define the limits of the system, the type of data required, the environmental effects to be considered, limitations, etc. The second phase, life cycle inventory assessment, takes a detailed look at the processes that occur within the system, considering the input and output flows (a general example is shown in Fig. 4.1). In the life cycle impact assessment, the information from these flows is used to establish product improvements, energy savings and emissions reduction, to name a few. Finally, in the interpretation, the results obtained are evaluated and validated. To achieve this, sensitivity analyses are usually carried out to test the quantitative effects on the impacts of the assumptions made throughout the study.

The life cycle assessment represents progress in evaluating the environmental impact of anthropogenic action on the environment. Thus, it has become a very popular tool for evaluating resource consumption and the environmental impact of products and services. It is, in fact, becoming an increasingly flexible, mature, comprehensive and transparent type of analysis.

These analyses are carried out using software that has access to a database with information on energy consumption, materials, chemicals, metals, agricultural aspects, waste management and transportation. However, in this database, the estimates of energy consumption for the extraction of certain elements and their derived environmental impacts are not very precise. This data also depends on the available technology since there is great variability between countries and practices. Arithmetic means are sometimes used, which do not always reflect reality in the case of mineral resources.

Assessment of abiotic resource depletion can be carried out using this life cycle assessment, however, conventional approaches do not adequately reflect the dispersion of minerals. This is because the assessments focus only on the part of the process

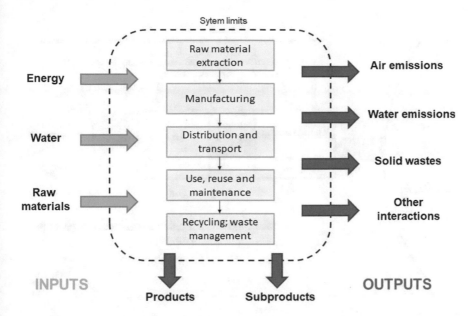

Fig. 4.1 Input and output flows typically considered in life cycle assessment

that goes from the cradle to the grave, focusing on the environmental impact associated with the products through inventories of materials, energy and environmental aspects. Next, the cycles of the minerals from the cradle to the grave are defined, then the analysis from the grave to the cradle will be introduced.

If we consider a cycle for each individual mineral, either from a biological or geological point of view, we can obtain enough information to understand the theory behind it (Fig. 4.2).

For millions of years, nature has formed and concentrated minerals in the deposits that we know today through different magmatic, hydrothermal and sedimentary processes. Such deposits represent the natural stock or the cradle. The existence of these deposits, with high concentrations of minerals with respect to the average concentration of the Earth's crust, makes it possible for society to avoid enormous amounts of energy when extracting them. Minerals are extracted from this cradle and undergo different processes (crushing, classification, concentration and metallurgy), which involves reducing the natural stock on Earth. Then, with the elements already concentrated from the minerals, products are created that later arrive on the market, satisfying the different needs for which they were created. As long as these products remain in circulation in society, the minerals they contain can be considered as stock in use.

At the end of the product's useful life, it becomes waste, either in the form of contamination or ending up in a landfill. In some cases, products will undergo a recycling process; costs related to mining and concentration would then either disappear

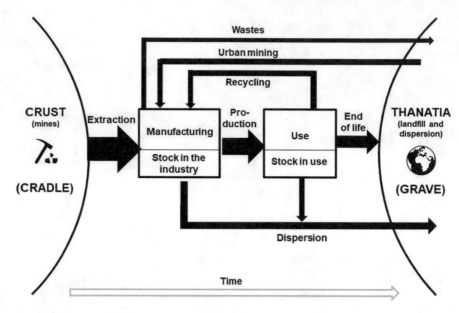

Fig. 4.2 The life cycle of materials (Valero & Valero, 2014)

or be reduced. Still, many products end up in landfills as they are virtually impossible to recycle and recover separately. Eventually, these elements will degrade and disperse to become part of the grave. In all phases, there is also a generation of waste that contributes to the degradation and dispersion of materials. Mathematically, this cycle can be expressed by the following Eq. (4.1):

$$input - output + recycling = stock\ changes + dispersion \qquad (4.1)$$

By extracting ore from a mine, a quantity of energy is spent in the form of fossil fuels and, on the other hand, the quality of the mine decreases. Its corresponding environmental footprint has two components: the impact of the extraction of the mineral, and the debt generated by causing the final dissipation of this mineral previously concentrated by nature. Each natural resource has an environmental cost associated with its dispersion, so any process that reduces this primary extraction is positive, be it for the replacement, recycling or improvements in the mining process's efficiency.

From the perspective of future generations, having natural resources concentrated in "warehouses" instead of in mines would be a better option, since environmental impact would become a thing of the past. Something similar happens with pyramids, already abandoned, or cathedrals, created for humanity's future enjoyment. The problem with mineral resources is that they are not stored, they disperse when used. Any resource that ends up in a state of dispersion, if combined with increased demand, can only be offset by recycling or more primary extraction. Every time human beings' resort to the use of non-renewable resources, without replacing them, it causes their irreversible loss. However, there is no way to assess what valuable materials for humanity have been lost forever, and this is because the scarcity and replacement cost of resources is rarely considered in conventional accounting methods.

Dispersion is the key to understanding the path of material resources and to developing new and more robust accounting methods, but the dispersion of materials has not been sufficiently included in economic analyses either; it has been considered more as a contamination problem than as an availability problem. Dispersion can only be considered by performing a material balance, as with energy balances (Eq. 4.2):

$$extraction + recycling - stock\ in\ society = dispersion \qquad (4.2)$$

Indeed, neither the current economic nor physical accounting systems are robust or detailed enough to consider mineral resources' dispersion. However, the energy, water and materials needed for mining, processing, manufacturing, use and recycling, and even materials that end up in landfills are accounted for in life cycle assessments using a cradle-to-grave approach. These are partial analyses, as they only consider one half of the problem.

To close the loop, cradle-to-grave analysis can be combined with grave-to-cradle analysis. For this to happen, we need to have a reference to be able to carry out the calculations. The only way to do this rigorously is by using the exergy and replacement costs of the different minerals, concepts that will be discussed later and that will allow us to take into account not only the quantity but also the quality of the extracted minerals.

As established, the results derived from the life cycle assessment depend on the chosen system limits, without there being absolute values for a given product or service. However, if the analysis is started from a hypothetical grave in which all mineral resources and fossil fuels have been exhausted, that is, Thanatia, a dead planet, this would serve as a limit and as a reference, necessary to be able to carry out the calculations for any given product. Theoretically, absolute values of life cycle assessment could be obtained by linking this type of analysis with the second law of thermodynamics. Therefore, Thanatia could be used as a starting point for the assessment of the depletion of abiotic resources.

Although Fig. 4.3 represents a general diagram, each specific substance can have its own cycle. The traditional analysis of the life cycle of materials ranges from the cradle (R#1) to the grave (R#0), through the stock in use (R#2 and R#3) and landfills (R#4). The complementary part, from the grave to the cradle, would be the path that should be followed from R#0, where the minerals appear totally dispersed in the Earth's crust, to the cradle (R#1), a planet where the minerals appear concentrated in mines. Each of these points on the road can be considered a reservoir. The size of each reservoir depends on the amount of mineral available (R#1), on the annual extraction rate (from R#1 to R#2), on the annual deposition rate (from R#3 to R#4), the annual recycling rate (from R#4 and R#3 to R#2) and the amount of loss associated with any process. In addition, each reservoir should have at least three properties: quality, chemical composition and concentration.

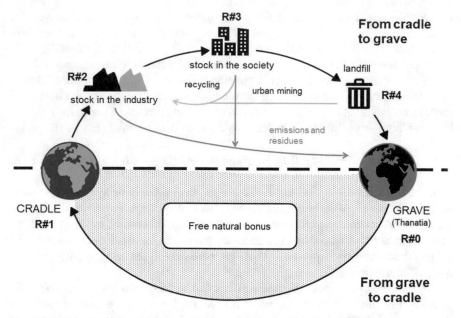

Fig. 4.3 Closing the life cycle of materials: the analysis of cradle-to-grave and grave-to-cradle

Based on this, two physical indicators can be proposed for each reservoir: exergy and exergy cost. Although these concepts are explained in detail later, it should be noted here that exergy is the contained energy measured in units of the highest energy quality, for example, equivalent kWh, while exergy cost is the cost measured in exergy units of all processes required to manufacture a product from zero. In other words, in the first case, exergy is an indicator that measures how far removed is something from its reference environment (in this case, Thanatia). In the second case, it measures the expenditure of exergy in the production process, once the limits of the analysis and the process, and its efficiency, have been defined. In this way, the loss of mineral wealth can be realistically quantified.

Once we know what our grave is going to be, that is, the limit of our analysis, we then need to define what its composition will be before we can describe what indicators will be used in the process of mineral reconcentration.

4.2   Thanatia

The word *Thanatia* comes from the Greek *Thanatos*, which means death, and represents a hypothetical geosphere where all concentrated materials have been extracted and dispersed throughout the crust, and where all fossil fuels have been consumed with subsequent effects on the atmosphere and hydrosphere. That all mineral deposits on the planet have been depleted does not mean that the planet has reached the end of its life—it only implies that mineral resources will no longer be available in their current form, but rather are dispersed. The second law of thermodynamics dictates that the entropy of the universe always tends to the maximum, that is, it tends towards thermal death. However, in Thanatia this is not exactly the case.

Let us compare the Earth with Venus, a planet much closer to the Sun, with an atmosphere composed of 95% CO_2. It is impossible for the Earth to reach such a state at least on a human time scale, since its temperature, not being so close to the Sun, will never reach those limits. At the relatively low temperatures that the Earth is exposed to, most of the components of the crust are in a solid state, one exception being water in a liquid state (essentially inert at room temperature). In water, the most important reaction is that of dissolving substances; in air, the most important processes are oxidation and erosion.

Since the appearance of life on Earth, the composition of the atmosphere has remained more or less constant, with 21% oxygen, 78% nitrogen and 1% composed of other minor gases such as argon, carbon dioxide or water vapour. Its distance from the Sun is a determining factor in preventing the entropic degradation suffered, for example, by Venus. At the same time, this distance from the Sun has allowed for the appearance of life on Earth, as well as the cycles of different elements such as carbon, nitrogen and water, including the cycles of materials, making our planet truly unique.

As long as the Sun continues to radiate its heat, it will continue to be the greatest source of energy that Earth receives to build and destroy natural resources. Of all the energy received from the Sun, about 0.023% is used for photosynthesis, and this small fraction is already enough to maintain the entire planet's biological envelope. Therefore, if the Sun continues to exist and there continues to be an internal heat source on Earth, the planet cannot achieve balance or thermodynamic death.

However, the greatest danger on Earth right now is humanity itself, since we are responsible for accelerating natural cycles, extracting mineral resources and consuming fossil fuels that have taken millions of years to generate, increasingly degrading the planet. Therefore, it is not so complicated to imagine a commercially dead planet as a possible end of the so-called Anthropocene, a hypothetical scenario where all the concentrated materials in mines have been extracted and dispersed, and where all fossil fuels have been consumed, leading to an increase in CO_2 in the atmosphere, with the consequent increase in planet temperature due to the greenhouse effect. Using this dead planet as a reference environment, each substance that is more concentrated or diluted, that is colder or hotter, that has a higher or lower chemical potential, pressure, height or speed, among many other properties, will have exergy.

Other planet models exist, such as Gaia or Medea (Lovelock, 1972; Ward, 2009). In the first of the models, Gaia, developed by James Lovelock and Lynn Margulis, the planet is considered as an organism capable of self-regulation, which contributes to the maintenance of those conditions necessary to maintain life on the planet. It is an environmental hypothesis in which the biosphere and the physical components of the Earth are integrated into a complex and interactive system and whose buffering capacity is gradually decreasing (Lovelock, 2006). This capacity decreases every time human beings contribute to the loss of biodiversity and forests, every time they consume fossil fuels, causing an increase in greenhouse gases, etc. In the case of the Medea model, named after Jason's wife and who out of spite killed her own children, life on Earth, instead of evolving symbiotically with its environment, tends to suicide. That is to say, life itself is the main threat to its own continuation. This builds on the past's mass extinctions when life reached such an expansion that it destroyed available resources until nothing was left. Unlike the Gaia model, in the case of Medea, there is no negative feedback, only a biocidal trend; no possibility of recovery, only absolute devastation, followed by the development of other species(s) taking advantage of the opportunity.

Both models predict an impending catastrophe, partially connecting them to the second law of thermodynamics. Any system tends to evolve until it reaches equilibrium with its environment, extracting and destroying all the exergy. In the case of Medea, if a natural species is a thermodynamic system, it will reproduce to such an extent that it will end up exhausting the ability of its environment to maintain it. This philosophy is thus even closer to Thanatia's than to Gaia's. The fundamental difference between these two models and Thanatia is that in the latter it is human beings who are directly causing their own extinction or at least the death of current civilisation, that is, it represents death through the exhaustion of their resources. Note also that neither Gaia nor Medea describes the composition and exhaustion of the Earth's mineral resources.

This crepuscular planet, Thanatia, would have an atmosphere, a hydrosphere and a continental crust with a specific composition, just like the planet Earth we all know today. Thanatia's atmosphere is marked by an increase in the concentration of CO_2, having burned all the fossil fuels that exist. It, therefore, presents a slightly different concentration than the current one. The atmosphere of our crepuscular planet, as evaluated, would have a CO_2 content of 683 ppm, an average temperature of 17 °C, that is, 3.7 °C above pre-industrial levels, a pressure of 1.021 bar and a composition in volume of 78.8% of N_2, 20.92% of O_2, 0.93% of Ar and 0.015% composed of other gases (Valero & Valero, 2014). In the case of the hydrosphere, the burning of fossil fuels and the increase in temperature would produce changes in the water content of the atmosphere or in the ice sheets, among others (Valero et al., 2011a). We know that 97.5% of the water on Earth is contained in the oceans in the form of salt water, the rest corresponding to fresh water. Of the latter, more than 90% is in the poles and Greenland and only 0.01% is in rivers and lakes. It is reasonable to state that, despite the blending of meltwater and fresh water, the composition of Thanatia's hydrosphere would be very similar to that of the oceans today.

It will be, therefore, in the continental crust of Thanatia where the changes will be greater than in the current crust. The solid part of the Earth is made up of well-differentiated layers, from the inside to the outside, the core, mantle and crust. The core is in turn divided into two parts, solid and liquid, and its composition is mainly iron and nickel. The mantle is next, made up of a layer of solid but plastic rock, rich in silica, aluminium and oxygen. The outermost layer is the crust, which can be divided into oceanic and continental crust. The oceanic crust is thinner and denser than the continental crust, being on average about 7 km thick and composed of mafic rocks, that is, rocks rich in magnesium and iron, such as basalt. The continental crust can reach 40 km in thickness; here we find any type of rock and mineral, serving as a reservoir for various mineral resources. However, only the upper part of this continental crust is accessible to humans using techniques available today.

Various estimates show that the mass of the upper part of the continental crust represents approximately half of the total continental crust, which would correspond to a volume of 6.55×10^5 km^3. According to estimates by the United States Geological Survey, there would be a total of 10^{17} kg of mineral resources in the crust (USGS, 2019). To this, we must add fossil fuels (both conventional and unconventional), in the order of an additional 10^{16} kg. As a whole, minerals and fossil fuels would barely represent 0.001% of the total mass of the upper continental crust. This figure already gives us an indication of how rare mineral deposits are in the crust itself. The thermodynamic properties of the upper continental crust were calculated and presented in Valero et al. (2013).

There are different models to estimate the chemical composition of the upper part of the continental crust (Clarke, 1889; Condie, 1993; Ronov & Yaroshevsky, 1969; Rudnick & Gao, 2004). On the other hand, there are also models on their mineralogical composition, although they are much more complex due to the heterogeneity of the minerals. We should note that, according to the International Association of Mineralogy, more than 5,000 mineral species are known in the crust, of which 150 are common and the rest range from infrequent to very rare. The first study of the crust's mineralogical composition was carried out by Grigor'ev (2007), based on over 3,000 rock samples, creating a list with the estimated concentration in the crust of the 265 most important mineral species.

If we take into account the tiny part that concentrated mineral deposits represent with respect to the continental crust, 0.001% as we have seen, it can be assumed without much margin of error that the average composition of the crepuscular crust is equivalent to the average crust of Earth. Starting from the mineralogical composition determined by Grigor'ev (2007), once the values were updated, other minerals that had not been taken into account thus far were added, and considering the crust as a closed system, we were able to determine the composition of the crust of Thanatia, our exhausted planet. The complete calculation methodology of how this concentration data was obtained for the different minerals can be found in (Valero & Valero, 2010; Valero et al., 2011b).

In this model of continental crust, there are 324 mineral species, 57 more than in Grigor'ev's model. Of these, 292 are minerals, the rest are minor elements that can appear to replace other major minerals, such as cerium, dysprosium, gadolinium, etc. The result is that the molecular weight of the upper part of the Earth's crust is 155.2 g/mol.

This model, as mentioned, has been obtained through empirical data, since it is impossible to explore the continental crust in such detail. That said, this composition of Thanatia's crepuscular crust will later allow us to assess the loss of Earth's mineral wealth from each of the 78 stable elements present in it.

Having access to a reference environment for carrying out exergy calculations is a fundamental question that has been studied by many authors previously (Ahrendts, 1977; Bosjankovic, 1963; Gaggioli & Petit, 1976; Sussman, 1979; van Gool, 1998).

To carry out calculations using exergy, we need to have this reference environment, since exergy is an exergy measure of how different something is from our environment. The greater the difference, the greater its exergy. Since this reference environment is associated with thermodynamic limits, it is usually considered a "dead" environment. This means it has reached equilibrium with its environment, it has lost all its exergy and, consequently, all capacity to produce work. In the case of mineral resources, the properties that are critical and must be analysed are the concentration and composition of their environment, which is what makes them valuable and allows us to analyse the loss of mineral wealth, since other properties, such as temperature, do not affect its quality.

Thanatia, which can be used as a basis for analysing this loss of mineral wealth, should not be confused with the reference environment defined by other authors, the most commonly used being that of Szargut (1989). That model allows us to carry out chemical exergy calculations, establishing the minimum work that would need to be done to obtain a resource with a specific composition from the reference environment, not from Thanatia.

Within a reference environment, reference substances must be divided into gases (in the atmosphere), solids (in the lithosphere) and dissolved substances (in sea water). On these terms, Thanatia resembles this reference environment, since it is also composed of an atmosphere, hydrosphere and continental crust. However, the conventional reference environment is usually composed of a maximum of about 85 chemical substances, with a single substance assigned to each chemical element. These substances vary between the different models of reference environments according to the authors' criteria. This does not happen in the Thanatia model, which, as we have seen, is made up of many more substances and approximates the reality of the Earth's crust.

Various authors have created reference environments based on different hypotheses to calculate the chemical exergy of the elements. One example is the Ahrendts reference environment (Ahrendts, 1977, 1980), which consisted of only the 15 most common elements on Earth, later updated by Diederichsen (1999) to 75 elements. They began from the hypothesis that these elements reacted with each other until they reached a hypothetical state of equilibrium. However, the time required to reach this balance is so high that it would not serve in the case of Thanatia, where the aim is to see the effects on human activity. Another widely used reference environment, mentioned above, is Szargut (1987, 1989), which is much closer to the natural environment since the substances present are the most stable and abundant. This model allows us to obtain a good representation of the products that would form in the interaction between the components of the natural environment and process residues. This model was also updated by Ranz (1999), developing a reference environment that contained the most stable minerals that appear in the upper part of the Earth's crust. These two models are more in line with the aims of Thanatia. Nonetheless, they cannot replace the model of the crepuscular crust since they only use a reference substance for a chemical element. If we take silicon as an example, this element appears in many substances naturally, and it is very unlikely that, on a human time scale, all of it will end up becoming a single reference substance, like SiO_2.

This is because, although possible from a thermodynamic point of view, the kinetics of the reactions would prevent that transformation.

Finally, we must not leave out the last fundamental difference between a reference environment and Thanatia. Only the chemical composition of the medium is taken into account. The exergy of mineral deposits increases exponentially as a function of their concentration, so the greater the difference between the concentration of a mineral in the mine and the crust, the greater the exergy that deposit has, the greater the difference. This aspect has been considered in Thanatia's crepuscular crust model but not in the other reference environments since it is not only necessary to know the chemical composition of the substances present but also their concentration. We can thus conclude that Thanatia, and what is usually called the reference environment, are two different concepts despite having certain points in common.

In summary, Thanatia is the description of a degraded technosphere with a specific composition of the atmosphere, hydrosphere and continental crust; it is a commercially dead planet where all mineral resources have been mined and fossil fuels burned. This scenario does not imply the end of life on Earth. In fact, we will only see this scenario play out if humanity fails to find alternative means for its development, which involve the closure of cycles, the use of other materials and renewable energy sources. Starting from Thanatia, it is therefore easy to assess the depletion of minerals by associating them with a cost, corresponding to the useful energy needed to concentrate the mineral components of any rock until its concentration is that of the current mineral deposit. In other words, moving from a dispersed environment such as *Thanatia*, the grave, to the current mines, the cradle, thus completing the conventional life cycle assessment.

Now that we are clear on what the Thanatia model consists of, we can carry out a simulation of the amount of energy required to extract minerals from this situation of absolute mineral dispersion.

4.3 Energy Needed to Extract Minerals from Thanatia

In this practical case, we will analyse the influence that the decrease of the ore grade has in the mines of gold on the final energy consumption and considering as a limit the concentration of this element in Thanatia.

To do this, let us first look at how gold could be extracted and concentrated from the crust and what factors need to be considered during the process. The processing of the mineral, from Thanatia to the generation of gold cathodes, is represented in a very simplified way in Fig. 4.4.

The material must first be moved to the treatment plants, then undergo the first stages of crushing and grinding, where it will be reduced to a certain particle size that allows for maximum recovery, which can be determined by mathematical calculations. This entails a specific associated energy expenditure. Once the material is ground, it must be pre-concentrated to attempt to separate as many ores as possible from the gangue. To achieve this, gravitational separation is usually employed, since

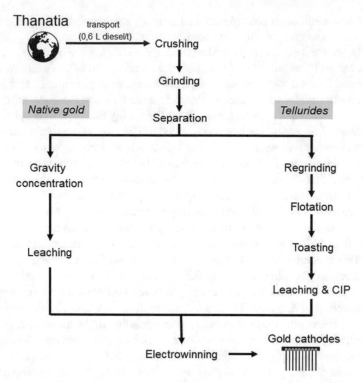

Fig. 4.4 Simplified process of the steps necessary to manufacture gold cathodes from Thanatia (Palacios et al., 2019)

gold is much denser than other elements. The process can then take two different paths depending on where that gold is present, whether in the form of native gold or contained in tellurides.

In the first case, for gold that is in the native state, the process is simpler since only gravimetric concentration and leaching are used. Leaching seeks to dissolve partially or totally a solid to recover the material contained in it; in this case the gold. A leaching liquid is usually used that varies depending on the properties of the element in question. This may be an acid, a salt, etc. The process is carried out at lower temperatures compared to other metallurgical processes. In the last phase of the gold concentration process, cyanidation is specifically used as a leaching technique, which consists of converting gold (insoluble in water) into soluble complexes by adding sodium cyanide.

In the second case, if gold appears in tellurides, the processing is somewhat more complex. First, the material must be reground and must undergo a flotation process. This process consists of concentrating the metal by injecting air bubbles into an aqueous medium. The mineral particles of interest remain attached to these air bubbles, thanks to the addition of chemical reagents that promote or decrease the hydrophobicity of metals. This metal attached to the bubbles remains floating in the foam, generating a pulp that is then collected. This resulting pulp is concentrated in successive tanks until the maximum possible recovery state is reached. It then undergoes a roasting process at temperatures between 600 and 700 °C, ensuring the telluride bonds are broken and the gold is released.

Once as much gold as possible has been recovered from the native gold or gold-bearing tellurides, the resulting material is subjected to an electrometallurgical process. This process consists of the extraction and refining of metals by an electric current, known as an electrolytic process. An electric current is passed through the aqueous solution where the gold we want to concentrate is; the present gold will thus slowly deposit on the cathode, forming the gold cathodes that is the final product (Norgate & Haque, 2012).

Considering the gold recovery process explained above, the energy consumption values of the different processes have been calculated from Thanatia to the obtaining of gold cathodes.

Table 4.1 shows the energy consumption, in GJ per ton of gold produced, of the different processes analysed, as well as the total energy consumption, being 6.2×10^7 GJ/t of gold. These calculations have been obtained simulating the processes described above with the HSC Chemistry software (Palacios et al., 2019).

For the calculations in Table 4.1, we have used an average gold grade of Thanatia, 1.28×10^{-3} g/t, a very low figure compared to current or even past time periods. Specifically, in Australia, the average gold grade in the late nineteenth century was almost 37 g/t (Mudd, 2007). The drastic decrease in the average ore grade in current mines is not only due to the fact that today technological advances have allowed mines with very low concentrations to be exploited profitably but also to the decrease in the amount of gold available in the mines due to intensive extraction. This decrease is in turn associated with an increase in energy consumption. In 1991, it was estimated that 172 GJ are needed to be consumed to obtain a kilogram of gold. Today, this figure is closer to 200 GJ (Calvo et al., 2016; Mudd, 2007). And if we compare these figures with those obtained here, we see that the energy that would be necessary to use to extract gold from Thanatia would be much greater.

As gold concentration decreases, as we move from the present value in the mines (2.72 g/t) towards a hypothetical future value, when the mines are almost exhausted and we move closer to Thanatia, the specific energy required will increase until we reach an exponential trend the closer we are to zero (Palacios et al., 2019).

On the other hand, this study has been carried out for specific particle size and the associated energy would increase considerably if the particle size decreased, the same thing that happens if the mineral grade is changed: this is represented in the different coloured bands shown in Fig. 4.5.

Table 4.1 Energy consumption of the different stages to obtain gold from Thanatia

Process	Energy consumption (GJ/t-Au)
Ore handling	1.62×10^7
Concentration	4.5×10^7
Total	6.17×10^7

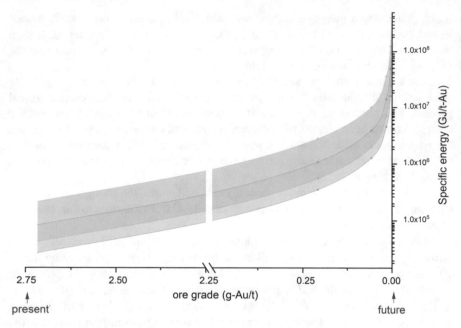

Fig. 4.5 Specific energy for the extraction of gold from Thanatia according to the ore grade (in grams of gold per ton)

In this way, we have verified that the extraction of gold from Thanatia would entail a very high energy consumption, from the extraction, transport, concentration and metallurgical treatments, which in turn is associated with the concentration of the mineral. If the concentration in the mines decreases, the energy associated with the extraction increases almost exponentially until we reach Thanatia as an absolute limit. Therefore, it is essential to optimise these processes to increase their efficiency and decrease energy consumption, as well as looking for other ways to recover the gold already on the market. Existing gold is found in jewellery, electronics and in other sectors. It is critical to promote the reuse of such a valuable metal instead of carrying out primary extraction, which will lead to the inevitable depletion of deposits over time.

The exercise carried out for gold can be performed for all the elements in the periodic table. However, this is not a trivial task, particularly due to the lack of available information. Alternatively, we can resort to theoretical methods to estimate what would be the costs of replacing this loss of mineral wealth from Thanatia. To this end, we will use exergy and the thermoeconomic analysis as our starting point. We will look more closely at the basic concepts of the theory in the following sections.

4.4 Thermoeconomics, Exergy and Exergy Cost

To evaluate the exergy of mineral resources, we must first be familiar with the laws of thermodynamics. The first law of thermodynamics, also known as the law of conservation of energy, dictates that energy cannot be created or destroyed, it can only be transformed. The second law of thermodynamics states that everything ends up degrading or dispersing, although it does not guarantee *when* this will happen. Thermoeconomics is the science that connects the saving of natural resources with physics and the economy through this second law of thermodynamics, the keyword here being *saving*, as the concept was created with a specific end purpose.

The term thermoeconomics was coined in the 1960s by Myron Tribus, Robert Evans and Yehia El Sayed when they used exergy to allocate water and energy costs in desalination plants (El-Sayed & Aplenc, 1970; Evans & Tribus, 1965; Tribus & Evans, 1962). Their original work consisted of relating the monetary cost of the physical flows, including the fuel and investment costs of the entire plant, with its thermodynamic utility, then called availability. Since then, many scientists have contributed to the development of this emerging scientific discipline.

Those who participate in the application of thermoeconomics try to avoid the waste of natural resources used for a certain purpose, one of the objectives being to universally avoid irreversibilities. Almost all human actions have associated consumption of physical resources that can be measured through the irreversibility they cause. On the other hand, natural resource accounting is associated with the concept of cost, a key concept in the economy; in this way, a link between physics and economics is established in thermoeconomics. Consequently, thermoeconomics is more interested in the cost of resources than in their price. This is because physics, in all its branches, aims to quantify and explain natural facts. Therefore, if one were to seek the physical cost of things, it would be necessary to search for tools that were as independent as possible from human will. Thus, it is obvious that money cannot, and should not, be a unit of measurement since it depends on many subjective factors such as market supply and demand. Quantifying nature in monetary terms opens a door to arbitrariness since nature is not actually selling us anything.

This is inexorably linked to the next question, what then should be the measuring unit of these physical costs? The answer can be found in the second law of thermodynamics. If the cost is a sacrifice of resources, and the resources consumed have already been consumed forever, this can serve as the basis for our physical accounting. Thermodynamics provides the necessary tools for the analysis of these costs: energy, entropy or exergy. The problem with the first, energy, is that it only measures quantities and not qualities: a kWh of work is not equivalent to the same amount of heat. Entropy is a somewhat better indicator since it measures the degree of molecular disorder of a system; coal, for example, has lower entropy than the gases generated after its combustion, and spontaneous processes only occur if the entropy increases. However, although entropy is sensitive to quality, the problem is that its units (energy/temperature) are not very practical. As with energy, it only gives information in one dimension. In fact, both properties, energy and entropy, are necessary to analyse a process in its entirety.

Although exergy is a one-dimensional property, it incorporates both types of information: it has energy dimensions and is sensitive to both the quantity and quality of the energy exchanged. Therefore, exergy is a suitable unit of measurement for measuring costs. As a concept, it participates in all the cost characteristics, is additive and can be calculated in all the operations that interact within any manufacturing process. It is a function of the quality and quantity of useful energy that can be used throughout the system; its most important contribution is its ability to objectify all physical manifestations in energy units, regardless of their economic value. In this way, any polluting product, resource, residue or emission can be objectively evaluated from an exergy perspective. Similarly, exergy considers all the physical manifestations that differentiate the system from its environment, be it height, speed, pressure, temperature, chemical composition, concentration, etc.

Exergy measures the minimum amount of useful energy required to form a system from its constituent elements found in a reference environment. Once that reference environment has been defined, the minimum thermodynamic cost or exergy of any material or energy flow can be calculated.

The exergy value of any resource implies the assumption that there has been no irreversibility involved in obtaining it. However, only in a thermodynamic utopia would processes be reversible, that is, they would evolve slowly enough to reach equilibrium and not be subject to friction or dissipation. In such a case, no additional exergy would be destroyed, and the manufacturing cost would simply be a reflection of the minimum thermodynamic cost (exergy).

Friction and dissipation are necessary for life; in the case of everyday processes, it is easy to see the need for light to see or friction to move. The conversion of electricity into a luminous flux and heat is a completely irreversible process; the same is true of friction, without which humans could not walk fully upright. For the Earth to be habitable requires the constant destruction of solar radiation. After all, life is an irreversible process, death being the absence of that irreversibility.

Biological processes are closer to reversibility than artificial ones; atmospheric CO_2 fixation through plants, for example, is carried out at room temperature using only sunlight. However, for humans to reduce CO_2 to pure carbon, or to synthesise hydrocarbons, a very complex technology with a very high associated energy consumption is required. On the other hand, many things that society appreciates have no value from a thermodynamic point of view, as is the case with stone sculptures, such as those of Michelangelo or Rodin, or even a ring made of gold. The exergy content of these items is practically zero. Since stone or gold appear in nature in the form of pure rocks and metals, their manipulation does not add additional exergy. This serves to clarify that the source of an object's value may not be related to its exergy content.

This is where we need to resort to exergy cost as opposed to exergy. The exergy cost can be defined as the sum of all the resources necessary to build a product from its components, expressed in exergy units and predetermined limits of analysis. This concept was originally introduced to international science in the 1980s (Szargut,

1987; Valero et al., 1986). The exergy cost value of a material is the sum of its exergy and all the irreversibilities that occur in its production process. The higher this exergy cost, the lower the efficiency of the production chain.

The concept of exergy cost belongs to the world of thermodynamics, but it also shares a large part of the conventional calculation methodologies of economics so that one can go from thermodynamic costs to thermoeconomic costs. This can be done simply by transforming exergy costs of the resources into resource prices and adding to each subsystem its annualised installation rate and maintenance costs. However, there is a difference between the two: exergy costs, unlike economic costs, relate physical measurements such as mass flows, pressure, temperature, and composition, with real irreversibilities that occur in the system. Thus, the basis of a general theory that improves systems and saves resources lies in systematically locating irreversibilities inside and outside the system and counteracting them with the money required to compensate them. Exergy cost-based accounting offers a broader scope and a clear view of the use and degradation of energy and other natural resources.

Costs are an emergent property since they cannot be found in the product as such; they cannot be measured as a physical magnitude of a flow or of a mass, composition or temperature. Instead, they depend on the structure of the system and appear because of the analysis; therefore, their calculation from physical data requires precise rules. The system also needs to be disaggregated when multiple products are produced simultaneously to avoid problems related to precise cost allocation.

Thermoeconomics is, therefore, a clear and unique way of connecting the measurement of physical loss, that is, irreversibility, with the loss of resources at the general level of the system, and also connecting it with other forms of reasoning, including economic ones. However, it should be considered that exergy only measures the number of times that a product is thermodynamically equivalent to another; it is a measure of the thermodynamic distinction of a system against a reference environment, not a measure of the value that can be given to things; it only evaluates the amount of energy used so that such a difference may exist. The exergy and the exergy cost serve to provide complementary, not universal, information on the footprint that humanity is leaving on Earth.

4.5   Exergoecology

Exergoecology derives from thermoeconomics and its objective is to evaluate the exergy of natural resources from a dispersed state, considering costs as a measure of resource sacrifice, regardless of the criteria used to quantify this loss (Valero, 1998).

Exergoecology can be divided into two main branches: physical hydronomics, which studies the exergy of water resources, including the alteration of the physical and biological properties of waters due to human activity, and physical geonomics, which studies the exergy of mineral and edaphic resources. Specifically, physical geonomics quantifies the amount of exergy that is saved by extracting mineral resources from mines, instead of the common rocks that would be the composition of a hypothetical planet that would have reached its maximum state of mineral exhaustion. This can help us not only to understand the energy we save by extracting minerals from mines, but also to quantify how degraded and dispersed natural wealth is and the rate of this degradation.

The fundamental difference between physical geonomics and other exergy approaches from the cradle to the grave is that in the latter case, only the exergy cost is taken into account from the beginning to the end of the process—everything related to extraction, profit, processing and use of the mineral. With physical geonomics the circle is closed, by means of the grave-to-cradle assessment, analysing the exergy and the exergy cost required to return that already dispersed mineral to the initial concentration conditions in the mine. Both approaches are important. The cradle-to-grave approach helps optimise the use of energy and materials throughout the life cycle of a particular product. In contrast, in the grave-to-cradle approach, the goal is conservation rather than efficiency.

4.6 Exergy of Mineral Resources

As seen in Chap. 3, a mineral deposit is a rarity within the Earth, which has been formed over time and with different geological processes. These deposits, therefore, have exergy since the elements are tens, hundreds or thousands of times more concentrated than their average concentration in the crust. In a mine, the areas with the highest concentrations are extracted, which tend to be more profitable so that gradually the concentration of the mineral in the mine decreases over time, causing the exergy difference between the mine and Thanatia to be reduced. When the limit is reached, that is, when the mine is completely depleted, this difference is equal to zero, having lost the "advantage" that nature had given us by previously concentrating these minerals.

Technically, the exergy that we will denote with the letter "B" is defined as the maximum amount of work that can be theoretically carried out by bringing a system into equilibrium with its environment through a sequence of reversible processes. It is therefore a measure of the potential a system must change if it is not in thermodynamic equilibrium with its immediate environment. It is in a dead state when the system has the same temperature, pressure, concentration and composition as its environment since reactions cannot take place and there is no potential or kinetic energy. This reference environment must have specific characteristics, intensive properties (such as height, speed, density, temperature, pressure and chemical potential of the substances in the medium) or extensive properties (such as volume).

Regarding this reference environment, all the materials have an exergy content that can be evaluated through reversible hypothetical processes that increase or decrease their intensive properties in a quasi-static way. Contrary to what happens with mass or energy, it is not conserved but is reduced every time the substance performs irreversible processes.

In the case of mineral resources, the pressure, temperature, speed or height with respect to the reference environment do not inform us of their intrinsic value. In fact, if only those properties were considered, a mineral would have an exergy of zero. Their composition, concentration and grindability are what make them useful for society and unique with respect to their environment.

The exergy of a mineral can be calculated as the sum of three factors: its chemical exergy, its concentration exergy and its comminution exergy. Chemical exergy (b_{ch}) is the minimum theoretical work that must be put into play to synthesise minerals with their corresponding specific composition from the reference environment (this would be that of Szargut, mentioned above). Concentration exergy (b_c) is the minimum theoretical work that must be put into play to concentrate minerals to a certain level, with respect to the dispersed state of Thanatia. Finally, comminution exergy (b_{com}) is the minimum theoretical work that must be done for the cohesion energy of the mineral components in Thanatia to reach that of the minerals in the mines.

With this in mind, the total exergy of a mineral resource i ($b_{t,i}$), where n is the number of moles of the substance considered, can be calculated with the following equation (Eq. 4.3).

$$B_{t,i} = n_i \cdot b_{ch,\,i} + n_i \cdot b_{c,i} + n_i \cdot b_{com,i} \qquad (4.3)$$

4.6.1 Chemical Exergy

The exergy involved in the process of forming a given mineral includes that of creating the compound from its individual elements and that of the cohesion of the molecules to form the crystal structure of the mineral. This exergy can be calculated using the chemical exergy of the mineral, which is equivalent to the minimum work that would need to be done to obtain the resource with a specific composition from the reference environment.

Technically, chemical exergy expresses the exergy of a substance at room temperature and pressure. It is defined as the maximum work that can be obtained when a specific substance is brought, through reversible processes, to the state of the reference substances in the reference environment. This hypothetical process, by means of isothermal reactions with the reference substances, would release this exergy. Therefore, to calculate chemical exergy ($b_{ch,i}$) we assume a standard ambient temperature, a standard ambient pressure and a standard concentration of the reference substances in the medium. Based on this data, we can calculate it by making an exergy balance of a reversible formation reaction (Eq. 4.4):

$$b_{ch,i} = \Delta G_{f,i} + \sum_j r_{j,i} \cdot b_{ch,j} \qquad (4.4)$$

where $\Delta G_{f,i}$ is the Gibbs free energy of the substance i; $r_{j,i}$ is the amount of moles of the element j per mole of substance i and $b_{ch,j}$ is the standard chemical exergy of element j contained in substance i.

Thus, the chemical exergy of any substance can be calculated by knowing the Gibbs free energy (normally tabulated) and the standard chemical exergy of its constituent elements. The value of this standard chemical exergy is calculated from a reference environment and Szargut's proposition (Szargut, 1989) is usually the most widely accepted. The values used in this book for the standard chemical exergy of each element appear in Table 4.2 and correspond to an update by Valero (2008) of the reference environment of Szargut (1989).

The conventional reference environment used in this section only serves to calculate the chemical exergy of minerals, for the rest of the components, which are the concentration exergy and the comminution exergy. This reference environment falls short as it only takes into account one species per element. Because of this, for the concentration and comminution exergy, Thanatia will be the reference environment used.

In the case of fossil fuels, chemical exergy is difficult to calculate due to its complexity. However, it has been shown in many cases that this chemical exergy can approximate its higher calorific value corresponding to each one with a margin of error no greater than 2%.

Table 4.3 shows the thermodynamic properties (HHV is the high heating value, ΔH_f^0 is the standard enthalpy, s_f^0 is the standard entropy and B_i^0, the exergy) of the different types of fossil fuels, considering the different types of coals and also a range for oil.

4.6.2 Concentration Exergy

The minerals appear mixed with each other in the rocks, which also combine with each other to form mineral deposits. Concentration exergy can be defined as the minimum theoretical work that must be put into play so that the minerals, not agglomerated by cohesion forces, are at a specific concentration with respect to the dispersed state of Thanatia. This is expressed by the following Eq. (4.5):

$$b_{c,i} = -RT^0 \left[lnx_i + \frac{(1 - x_i)}{x_i} \cdot \ln(1 - x_i) \right] \tag{4.5}$$

where R is the universal gas constant, T^0 the ambient temperature and x_i the concentration of the mineral. The concentration exergy difference between the mineral concentration in the mine (x_m) and the average concentration in the Earth's crust (x_c) is therefore the minimum amount of exergy that nature virtually invested in concentrating minerals from the dispersed state in Thanatia to the mine (Eq. 4.6).

Table 4.2 Standard chemical exergy of the different elements (Valero, 2008)

Element	$b_{ch,i}$	Element	$b_{ch,i}$	Element	$b_{ch,i}$
Ag	69.7	Hf	1,061.3	Rb	388.8
Al	794.3	Hg	114.8	Re	561.3
Ar	11.7	Ho	979.3	Rh	183
As	494.1	I_2	175	Ru	315.2
Au	51.5	In	437.4	S	607.3
B	628.6	Ir	256.1	Sb	437.1
Ba	765.5	K	366.5	Sc	923.8
Be	602.6	Kr	34.4	Se	346.7
Bi	274.8	La	994.3	Si	854.2
Br_2	101.1	Li	392.9	Sm	993.9
C	410.3	Lu	946.6	Sn	547.6
Ca	723.8	Mg	629.6	Sr	758.5
Cd	293.2	Mn	484.6	Ta	974.8
Ce	1,054.2	Mo	730.5	Tb	999
Cl_2	124.2	N_2	0.7	Te	326.4
Co	308.9	Na	336.6	Th	1,214.5
Cr	584.4	Nb	900.2	Ti	904.4
Cs	404.5	Nd	969.8	Tl	193.8
Cu	134	Ne	27.2	Tm	952.5
Dy	974.9	Ni	232.5	U	1,196.1
Er	973	O_2	4	V	721.5
Eu	1,003.9	Os	370.8	W	827.7
F_2	556.1	P	861.6	Xe	40.3
Fe	376.8	Pb	232.2	Y	966.3
Ga	514.6	Pd	145.7	Yb	944.9
Gd	969.9	Pr	963.8	Zn	339
Ge	556.5	Pt	146.5	Zr	1,077.4
H_2	236.1	Pu	1,099.7		
He	30.4	Ra	825.8		

$$\Delta b_c = b_c(x_m) - b_c(x_c) \tag{4.6}$$

The exergy of a mineral increases as its concentration increases. This growth is not linear since, according to the second law of thermodynamics, the effort required to extract a mineral from a mine follows an inverse logarithmic pattern in relation to the ore grade. That is, as the ore grade decreases (the concentration of the mineral), the energy needed to extract that mineral tends to infinity. Therefore, it is this component

Table 4.3 Thermodynamic properties of fossil fuels (Valero & Valero, 2012a)

Fuel	HHV	ΔH_f^0	s_f^0	B_i^0
	kJ/kg	kJ/kg	kJ/kg K	kJ/kg
Anthracite	30,675	−136.2	0.9	31,624
Bituminous coal	28,241	−757.7	1.1	29,047
Subbituminous coal	23,590	−1,125.0	1.0	24,276
Lignite	16,400	−662.7	0.8	17,351
Fuel–oil 1	46,365	−622.1	2.8	46,259
Fuel–oil 2	45,509	−763.7	2.7	45,517
Fuel–oil 3	43,920	−1,279.1	2.6	44,002
Natural gas[a]	42,110	−3,117.4	8.6	39,394[a]

[a]Values of B_i^0 of natural gas in kJ/Nm^3

that makes the exergy of a mineral help give a more realistic vision than the mass (Wall, 1977).

Using the different equations in this section, the exergy of the concentration of different elements can be calculated (Table 4.4).

Fossil fuels are a type of mineral resource, and therefore also have chemical and concentration exergy that can be calculated with the same equations mentioned above. However, the quality of fossil fuels practically does not decrease with the mixture of its different carbon components, as it does among minerals—it remains practically constant. Therefore, the concentration exergy in the case of fossil fuels is not relevant and it is not necessary to take it into account to calculate its exergy—only the chemical exergy is required.

In addition to the concentration exergy, we must also consider the forces involved in the crystalline formation, the commutation exergy. This will be discussed below.

Table 4.4 Exergy of concentration of different elements (Stanek et al., 2017)

Element	x_c	x_m	M	$b_c(x_c)$	$b_c(x_m)$	Δb_c
	kg/kg	kg/kg	kg/kmol	kJ/kg	kJ/kg	kJ/kg
Fe	9.66×10^{-4}	7.30×10^{-1}	159.69	123.22	12.41	110.80
Cu	6.64×10^{-5}	1.67×10^{-2}	183.52	143.37	68.60	74.77
Pb	6.67×10^{-6}	2.37×10^{-2}	239.28	133.75	48.99	84.76
Zn	9.96×10^{-5}	6.05×10^{-2}	97.44	259.71	95.97	163.75
Au	1.28×10^{-9}	2.24×10^{-6}	196.97	270.14	176.19	93.95
Al	1.38×10^{-3}	7.03×10^{-1}	156.00	120.46	13.73	106.73
Cr	1.98×10^{-4}	6.37×10^{-1}	223.83	105.46	11.39	94.07

4.6.3 Comminution Exergy

If the main characteristics of a mineral resource were only its composition and its concentration, the concentration exergy equation (Eq. 4.5) would be sufficient to calculate the exergy necessary to concentrate a substance from Thanatia to the conditions in which it appears in the mines. However, this equation is only valid for ideal gas mixtures where the collision between the molecules is of an elastic type. In reality, minerals also have cohesion energy, which prevents them from spontaneously turning into a gaseous substance. This energy is strong enough that the minerals have to undergo physical comminution processes such as crushing and grinding to extract the valuable elements. Therefore, we must consider the comminution exergy in calculations, defined as the minimum exergy required to join solids from a dispersed state to a more cohesive state.

The energy required to separate a solid particle into smaller ones depends on different physical factors, such as hardness, surface, as well as depending on mineralogical composition and the distribution of the mineral in the rock. There are at least three semi-empirical models that link particle size with the energy required for crushing and grinding, known as grinding laws: Kick's Law, Bond's Law and Von Rittinger's Law, and they can be generally expressed with the Walker equation. From these laws we can infer that the energy required for grinding increases exponentially as the particle size decreases and they explain the current costs, but not the minimum costs or exergy required. Thus, in this section, we will look in detail at the required formula to calculate the comminution exergy applied to large rocks in Thanatia (Valero & Valero, 2012a, 2012b).

To obtain the comminution exergy of a certain rock in a mine, the initial state of fragmentation must first be defined, characterised by its mean geometric size (d_M). These fragments will be composed of both gangue and ore and will have a specific surface energy per unit mass. At the same time, we need to assume that the size of the rock in Thanatia is d_Θ, and that its surface area per unit volume is negligible when compared to the fragments contained in the mine. Therefore, the comminution exergy can be defined as the exergy saved by having the rock fragmented (d_M) instead of in Thanatia (d_Θ). The formula for calculating the comminution exergy is as follows (Eq. 4.7):

$$b_{com} = \Delta A_v \cdot \frac{\gamma}{\rho} = 6F_r \cdot \left[\frac{1}{d_M} - \frac{1}{d_\Theta} \right] \approx 6 \frac{F_r}{d_M} \quad \text{when } d_M << d_\Theta \qquad (4.7)$$

where γ is the surface energy (J/m^2), F_r is the roughness factor of the surface and ρ is the density (g/cm^3). The corresponding values of each variable for the different minerals are tabulated (Tavares & King, 1998; Tromans & Meech, 2002).

For exergy calculations, a comminution process is considered ideal if no loss of any kind occurs, neither heat nor kinetic energy of the fragments, since the compression load is used only to generate new surfaces.

As an example, following Eq. 4.7, the comminution exergy of a 100 m side galena fragment would be:

$$b_{com} = 6 \cdot 1.256 \cdot 1.868 \left[J/m^2 \right] / 7.40 \left[10^3 \text{ kg/m}^3 \right] / 100 \left[10^{-6} \text{ m} \right] = 19.02 \text{ J/kg}$$

This value can be compared with chemical exergy and concentration exergy, being $b_{ch} = 3{,}104$ kJ/kg (with a molecular weight of 239.27 g/mol) and $b_c = 80$ kJ/kg, specifically. Therefore, galena's comminution exergy is several orders of magnitude lower than its chemical exergy and almost 4,000 times less than its concentration exergy.

This result, however, can lead to incorrect conclusions, since the comminution exergy is proportional to $1/d_M$ and the smaller the particle size, the greater the exergy necessary to reduce the size of a certain sample. Therefore, grinding to a size of one micron will require a thousand times more energy than reducing the sample to a size of one millimetre. To ensure that all the atoms in a crystal structure have separated, the sample should be ground to nanometre size (10^{-9} m). In other words, due to its behaviour, comminution is a very energy-intensive process when it comes to finely grinding a material, but it is not as relevant when evaluating Earth's mineral wealth.

For this reason, since the current continental crust is fractured, the materials that appear naturally already pre-fragmented can be considered as an exergy bonus that nature provides us free of charge. However, as has been seen in the case of galena specifically, if viewed from a global point of view this bonus is not very significant in comparison to chemical exergy and concentration exergy. Therefore, the total exergy of a particular mineral can approximate the sum of its chemical exergy and its concentration exergy only (Eq. 4.8).

$$B_{t,i} = n_i \cdot b_{ch, i} + n_i \cdot b_{c,i} + n_i \cdot b_{com,i} \approx n_i \cdot b_{ch, i} + n_i \cdot b_{c,i} = B_{ch,i} + B_{c,i}$$
$$(4.8)$$

4.6.4 Exergy Contained in the Planet's Mineral Resources

To calculate the exergy contained in minerals, we must consider both the chemical exergy and the concentration exergy, since, as we have seen, the comminution exergy is practically insignificant in comparison.

This section shows the minimum exergy (B_t) for extraction in 2018 of the reserves and resources of a selection of the most frequent minerals. To achieve this, we have used data from different sources (Calvo et al., 2017a; USGS, 2019).

The total exergy of the selected minerals extracted in 2018 (25.8 Mtoe) is significantly small when compared to the 1.7 Gtoe or the 10.8 Gtoe that are available in the reserves and resources, respectively, of these same minerals (Table 4.5).

Let us now look at the case of fossil fuels. As explained above, the exergy of fossil fuels is closely linked to their high heating value, and the exergy contained in them

Table 4.5 Exergy content of production in 2018 of the reserves and resources of some of the most frequent minerals. Data in ktoe

Element	Ore	Exergy (ktep)		
		Extraction	Stockpiles	Resources
Aluminium (bauxite)	Gibbsite	857.2	80,009.0	214,309.8
Antimony	Stibnite	25.1	359.0	771.8
Arsenic	Arsenopyrite	7.4	229.6	2,338.4
Barium	Baryte	36.8	1,473.9	7,757.4
Beryllium	Beryl	<0.01	0.1	0.3
Bismuth	Bismuthinite	1.7	38.8	71.3
Cadmium	Greenockite	3.3	61.7	755.3
Chromium	Chromite	830.6	11,074.6	276,864.6
Cobalt	Linnaeite	33.6	1,703.6	34,792.7
Copper	Chalcopyrite	4,220.9	144,717.7	422,093.2
Fluorspar	Fluorite	243.2	10,484.5	222,271.2
Gallium	Ga in bauxite	<0.01	0.9	176.6
Germanium	Ge in sulphides	<0.01	2.3	81.9
Gold	Native gold	<0.01	0.5	0.8
Graphite	Graphite	785.6	194,293.8	675,804.7
Gypsum	Gypsum	828.1	–	–
Indium	In (in Cu, Pb, Zn ores)	0.1	–	4.5
Iron	Hematite	13,129.0	997,800.3	4,201,264.2
Lead	Galena	335.6	6,787.8	152,535.3
Limestone	Calcite	2,271.4	–	–
Lithium	Li in brines	13.2	2,171.0	6,125.4
Magnesium	Magnesite	391.1	32,367.2	161,835.8
Manganese	Pyrolusite	232.7	8,014.6	13,314.5
Mercury	Cinnabar	0.2	6.7	43.1
Molybdenum	Molybdenite	75.1	2,754.1	3,505.2
Niobium	Columbite	0.9	57.1	–
Niobium	Garnierite & Pentlandite	199.7	267.2	439.7
Phosphate rock	Fluorapatite	433.6	110,813.2	481,796.6
Platinum group elements	Cooperite	<0.01	6.0	9.1
Potassium	Sylvite	663.1	67,888.8	3,947,022.1
Silver	Argentite	1.9	39.7	91.0
Strontium	Celestite	1.7	45.0	6,611.6
Tantalum	Tantalite	<0.01	0.9	–
Tellurium	Te (in Cu ore)	<0.01	1.6	3.1

(continued)

Table 4.5   (continued)

Element	Ore	Exergy (ktep)		
		Extraction	Stockpiles	Resources
Tin	Cassiterite	2.5	39.0	618.8
Titanium	Rutile & ilmenite	151.3	15,905.1	–
Tungsten	Scheelite	1.1	43.8	–
Vanadium	V in several ores	25.2	5,168.3	21,706.8
Zinc	Sphalerite	15.3	37,309.2	354,437.4
Zirconium	Zircon	5.0	252.2	201.7
Total		25,808.2	1,694,879.4	10,855,218.7

can approximate their chemical exergy. In the case of oil, the error of considering the exergy content equal to the high heating value is only 0.26%, in the case of natural gas this value is somewhat higher, 6.5%. Finally, in the case of coal, depending on which type of coal is taken into account, the error varies between 3 and 6%.

To carry out the calculations, we have taken as reference the proven resource values and average composition for each fossil fuel: oil, natural gas and coal (BP, 2019; Valero & Valero, 2014).

Table 4.6 shows the results and the total exergy content of extraction and proven fossil fuel reserves by region and in total across the world. The exergy contained in the proven reserves of oil, natural gas and coal is 244, 185 and 689 Gtoe, respectively. The value for coal is almost three times that of oil and four times that of natural gas. Compared to the exergy content of production in 2018, there are differences of two orders of magnitude. Furthermore, the exergy associated with non-energy mineral resources (10.8 Gtoe) is less than 1% of that of fossil fuels (1,118 Gtoe).

The figures obtained show us that exergy is insufficient in the fair evaluation of minerals since common sense dictates that all mineral resources including gold, platinum, copper, aluminium, etc. are not less than 1% of fossil fuels. From a human

Table 4.6   Exergy content of the extraction (E) and the proven reserves (PRes) of fossil fuels by region (BP, 2019). Data in Mtoe

	Oil		Natural gas		Coal	
	E	PRes	E	PRes	E	PRes
North America	1,027	35,423	906	13,060	401	179,025
Central and South America	335	51,071	152	7,675	60	8,973
Europe and Eurasia	872	21,502	930	62,535	446	195,260
Middle East	1,490	113,216	591	70,827	0.7	869
Africa	389	16,585	203	13,534	156	9,531
Asia Pacific	362	6,345	543	17,007	2,853	295,667
Total	4,475	244,142	3,325	184,638	3,917	689,325

perspective, the value of minerals is more closely linked to their cost of extraction. A highly concentrated and abundant mineral in the crust, such as iron, has a high exergy value but a low extraction cost. A scarcer mineral, such as gold, has a low exergy value but a high exergy extraction cost.

Thermodynamics can help us predict behaviours, but in this case, the energy required to recover these minerals is several orders of magnitude higher than the thermodynamics of reversible processes dictate. In fact, mixing and separating are highly irreversible processes—when for example sugar and salt are mixed, the energy released in that process is almost imperceptible. If this mixing process were reversible, the energy required to separate the sugar from the salt should be the same. However, in practice, this is not the case.

Instead of using only exergy to assess mineral wealth, we will use exergy replacement costs, since they are more closely related to prices than exergy. Next, we will explain how we can obtain the mineral exergy replacement costs.

4.7 Exergy Replacement Costs

Exergy replacement costs (ERC) give us an idea of the amount of energy that would be necessary, using conventional technologies, to achieve concentrations in mineral deposits, from their concentration and composition conditions, from Thanatia. In other words, we will assess the life cycle from the grave to the cradle, rather than from the cradle to the grave. As we have seen, the exergy of a mineral has two fundamental components, excluding comminution: chemical exergy and concentration exergy, the sum being the total exergy. The chemical exergy is obtained from a reference environment and for that of concentration we need to consider the mineral grade in the crust and the mines.

Exergy replacement costs are defined as the total exergy needed to extract and concentrate the minerals from Thanatia, using currently available technology. Therefore, they are going to be a function of the mineral being considered, of its mining grade, of the separation and extraction technologies, and its associated energy consumption, which in turn may vary over time. The exergy difference between Thanatia and the mine will increase as the quality of the mine (its concentration) increases so that as the mineral deposit is depleted, the exergy will be reduced until it reaches zero.

To calculate exergy replacement costs ($b_{t,i}^*$), we first need to know the chemical costs related to the chemical production processes of the substance ($k_{ch} \cdot b_{ch,\, i}$), and the concentration costs, related to the concentration process ($k_c \cdot b_{c,i}$), as defined by the following Eq. (4.9):

$$b_{t,i}^* = k_{ch} \cdot b_{ch,\, i} + k_c \cdot b_{c,i} = b_{ch}^* + b_c^* \qquad (4.9)$$

The chemical exergy cost (b_{ch}^*) of a resource is associated with the synthesis of the mineral if it does not exist as such in the reference environment. However, since all the mineral substances are present in Thanatia, the synthesis process would disappear, and it would only be necessary to concentrate the minerals already present. In short, we will only take into account the exergy replacement costs associated with concentration (b_c^*).

The dimensionless variable k represents the unit exergy cost of a mineral and is defined as the relationship between the energy invested in the actual process of extraction and concentration of the mineral from the mine (x_m) up to the pre-casting and refining phase (x_r) and the minimum theoretical energy (exergy) necessary to carry out the same process (Eq. 4.10).

$$k = \frac{E(x_m \rightarrow x_r)}{\Delta b_{x_m \rightarrow x_r}} \tag{4.10}$$

Variable k is a measure of the irreversibility of processes, or in other words, of our technological ignorance. In an ideal (or reversible) process, k would equal 1 and exergy costs would equal exergy. The greater our technological ignorance, the greater k is and the greater the energy efforts necessary to carry out a certain process are.

Since the energy required for the mining process is a function of the ore grade and the technology used, so is this unit exergy cost. The lower the grade of the mineral, the greater the energy required to extract it. Technological developments tend to improve the efficiency of the mining process and, therefore, decrease energy consumption. In other words, the unit exergy cost depends on the mineral grade (x) and the time (t) (Eq. 4.11).

$$k = k(x, t) \tag{4.11}$$

This variable, k, can only be defined in this way for past events and for each specific mineral, so it is difficult to extrapolate it for future situations since the technological changes that may appear cannot be predicted. This variable is also not constant: the applied technology may change depending on the concentration of the mineral and depending on the type of mine (whether it is open-pit or underground, the particle size obtained after crushing, the depth the mineral is located at, etc.). However, considering the learning curves of the different mining extraction technologies, we will consider in this case that there will be no substantial advances that modify this unit exergy cost significantly so that it will only be influenced by ore grade variations.

Taking into account these limitations and the information available from the mining sector, which is typically rather scarce, it has been assumed that the technology used is the same in the extraction and refining process, from x_m to x_r, as in the reconcentration process from Thanatia up to the current conditions in the mines, that is, from x_c to x_m. In this way, the relationship between energy consumption and mineral grade for different minerals has been analysed and their unit exergy cost is calculated. Subsequently, it has been extrapolated to the concentration values presented by each mineral in Thanatia.

As an example, we will calculate the exergy replacement cost for copper. First, we need to have access to real data on energy consumption in mines based on ore grade (from x_m to x_r). This information can be obtained from bibliographic references, reports from mining companies, etc. In parallel, the theoretical exergy of that same process can be known using Eq. 4.6, when x is equal to x_m and x_r. Finally, the unit exergy cost can be calculated using Eq. 4.10 as a function of the mineral grade, extrapolating the results to the values of x_c, a figure that will be used to calculate the exergy replacement costs.

Copper is always associated with various metals, which are usually nickel, molybdenum or elements of the platinum group. In almost all the mineral deposits of copper that are exploited, this element usually appears associated with sulphur, in the form of chalcopyrite ($CuFeS_2$). This mineral has a concentration in the crepuscular crust (x_c) of 0.0000664 g/g. Considering the average ore grade in mines (x_m) is 0.01674 g/g (USGS, 1986), we can calculate the value of the concentration exergy, Δb_c from x_c to x_m, where R = 8.314 kJ/kmol K and $T^0 = 290.15$ K:

$$\Delta b_{c(x_c \to x_m)} = -RT^0\left[lnx_c + \frac{(1-x_c)}{x_c} \cdot \ln(1-x_c)\right] - \left(-RT^0\left[lnx_m + \frac{(1-x_m)}{x_m} \cdot \ln(1-x_m)\right]\right)$$
$$= 0.0728 \text{ GJ/t}$$

Therefore, the concentration exergy of copper would be 0.0728 GJ per ton of mineral, or 0.21 GJ per ton of copper, since in chalcopyrite there are 0.346 g of copper per gram of mineral. In literature, we can find values of $x_r = 0.82$ g chalcopyrite/g (Kennecott Utah Copper Corporation, 2004). Therefore, repeating the corresponding operations, the concentration exergy of the crepuscular crust at the moment before refining is $\Delta b_c(x_c \to x_r) = 0.13$ GJ/t.

Information regarding energy consumption according to the copper grade in mines has been obtained from different bibliographic sources (Calvo et al., 2016; Mudd, 2010); a summary of this data can be seen in Fig. 4.6.

From this figure, the energy consumption trend can be calculated with the following equation: $y = 25.65 x_m^{-0.366}$ GJ/t.

This means that when the ore grade reaches the grade in Thanatia, that is, when $x_m = x_c$, the energy will be:

$$E(x_c \to x_r) = 528 \text{ GJ/t}$$
$$k(x_c) = \frac{E(x_c \to x_r)}{\Delta b_{c\,(x_c \to x_r)}} = \frac{528}{0.13} = 1,387.4$$

Finally, the exergy replacement costs, that is, the exergy cost necessary to concentrate the mineral from Thanatia to the initial concentration conditions in the mines, from x_c to x_m, is:

$$b_c^* = k(x_c) \cdot \Delta b_{c\,(x_c \to x_m)} = 1,387.43 \cdot 0.21 = 291.70 \text{ GJ/t}$$

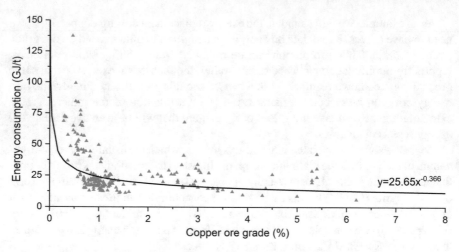

Fig. 4.6 Energy required to extract copper based on the ore grade (Calvo et al., 2016)

This calculation process, however, cannot be carried out for all the elements due to the lack of available information on energy consumption according to the mining grade. A general expression is thus proposed for an exponential curve that is used to calculate this consumption based on the mineral grade. According to the study of the data available for some elements (Valero et al., 2013), the equation is as follows, where the values can range from x^{-02} to x^{-09} (Eq. 4.12):

$$E(x_m) = A \cdot x_m^{-05} \tag{4.12}$$

The coefficient A is specific for each mineral. The average grade of the mineral in the mines is usually known, as well as the energy required to extract and concentrate the ore at a certain grade $E(x_m)$. Although it is an approximation, it matches the actual mining behaviour better than other equations that can be found in the bibliography (Chapman & Roberts, 1983).

Let us now see how, through exergy replacement costs, we can assign a value to resources that is more in line with the social perception of value than simply using their mass.

4.7.1 Exergy Replacement Costs as an Indicator of the Physical Value of Resources

The exergy replacement costs can be a very useful indicator that, unlike mass, considers the quality of materials as well as their quantity. If we take as an example the primary mineral extraction of 2018, we can carry out a comparison between the tons produced and its equivalent in millions of tons of oil equivalent (Mtoe) considering the replacement costs of each of the elements.

The left part of Fig. 4.7 represents the extraction of different minerals and fossil fuels, ordered from highest to lowest by weight, and on a logarithmic scale.

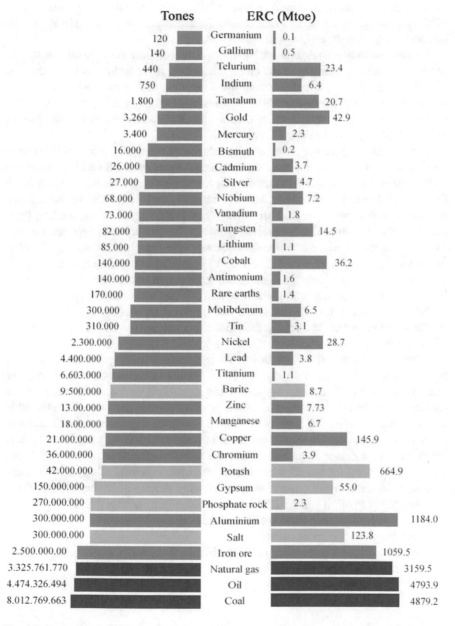

Fig. 4.7 Extraction in 2018, in tons (left) and in exergy replacement costs, in Mtoe (right) of the main metallic (green) and industrial (pink) minerals. *Data sources* USGS

If we were to add up the total extraction of iron and gold in tons for the year in question, we would see that the extraction corresponding to gold only represents 0.00013% of the total, a practically negligible figure. Is this figure significant? Considering that gold has an average price of $48 per gram and iron an average price of $70 per ton, carrying out this sum in tons does not seem very equitable and would be like adding very expensive pears to apples.

We have already seen that the replacement costs consider not only the quantity of material but also its scarcity in the crust. Therefore, if we add again the extraction of iron and gold in replacement costs, the final weight of gold would be 4%; this still seems low, but it is almost 30,000 times more than in tons. In this way, the importance of gold compared to iron would be much better reflected, and now we would be adding apples to apples.

The tons and replacement costs of minerals can be compared in turn with those of fossil fuels. Let us recall that the thermodynamic value of fossil fuel is represented by its chemical exergy, which in turn can approximate its high heating value. A fossil fuel, once burned, is lost forever, turning into CO_2 and water. Unlike fossil fuels, minerals are not lost when they are used—they eventually disperse, making them inaccessible for reuse. The mineral wealth of a deposit is being depleted for two reasons. The most evident is due to the decrease in the amount of resource available in the mine when it is extracted. The second is related to the decrease in the grade of the mine, which will imply a higher consumption per ton extracted as the deposit is mined. This dispersion component, which we can evaluate through the exergy replacement costs, is the one that is not being taken into account in the accounting for the use of resources, and yet it is essential to understand the loss of mineral wealth of a country, a region or the world.

What would happen if we took that factor into account in the global accounting for the loss of mineral resources in a given year? Figure 4.8 answers that question. The extraction (in tons) of oil, natural gas, coal and minerals is represented in the circle on the left, where the relative weight of each type of resource is fairly equal. On the other hand, when expressing this same extraction data considering the exergy replacement costs, these almost double their importance in weight. This is very significant since, although at first glance it may seem that the primary extraction of minerals is not very high, in 2018, its relative importance compared to fossil fuels almost doubles. In other words, the loss of mineral wealth in exergy terms was equivalent to nothing less than the sum of oil and coal. This is the vast loss of wealth that takes place on our planet year after year since we do not recover the raw materials once they are used and irreparably disperse them.

Next, we will see a practical application of how these replacement costs can help us in the mining industry.

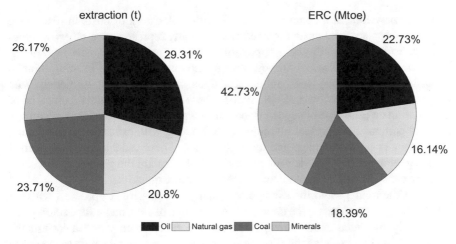

Fig. 4.8 Extraction in 2018 in mass (left) and in ERC (right) of minerals and fossil fuels. *Data sources* BP and USGS

4.7.2 Allocation of Costs in Mining and Metallurgy

As already explained, in the case of mining, it is rare that only one element is extracted from a mine, especially in the case of metals. In many copper, lead or zinc mines, there are an infinity of elements that appear in very small quantities, but whose recovery produces greater economic benefits than those of the main element. It should be borne in mind that the extraction of these minority elements will also depend to a great extent on existing recovery technologies, which may or may not make the process profitable. It is at this point, at the time to distribute the extraction and processing costs between each certain element, where problems begin to arise.

Traditionally, the life cycle assessment performs the allocation of costs between products based either on the weight of each (in kilograms or tons) or on the benefit obtained, that is, its sale price. Both approaches have certain disadvantages since in many cases the tons of primary elements and by-products do not even have the same order of magnitude, making comparisons meaningless. In the case of allocation based on sales price, it also incorporates a large number of uncertainties and subjective factors that do not represent a good distribution of these extraction costs. A recent example of volatility in the price of metals is that of rhodium, which has doubled in just one year, exceeding 140 euros per gram, a price three times higher than that of gold (USGS, 2019). This rise is fundamental since this element is used in catalysts so that vehicles pollute less and comply with environmental standards, hence the increase in recent demand.

To assign the extraction costs to each element, one option is to use the exergy as a unit of measurement, and specifically the exergy replacement costs to take into account the relative scarcity of each mineral and the energy intensity associated with its extraction.

Let us now look at a comparison between the allocation of costs in tons, price and exergy replacement costs for a series of globally representative reservoir types in order to test the validity and applicability of our approach. In the late 1980s, the United States Geological Survey, after an exhaustive study of existing mineral deposits throughout the planet, published a document with almost 40 models, with the corresponding subtypes (USGS, 1986). These descriptive models include basic information on each type of deposit (types of rocks that appear, textures, age, mineralogical composition, present alterations, etc.), along with examples and, in many cases, figures on the tonnage and average grades. These models are fundamental from a geological point of view since they not only explain the genesis of a specific site, but can also help to locate new similar sites more quickly.

Using the tonnage and the average ore grade, it is possible to proceed with the allocation of costs based on the tonnage, the sale price of each metal for certain years and, finally, by means of the corresponding exergy replacement costs of each metal.

In the first case, in which the allocation of costs is carried out through tonnage, the greatest impact will correspond to those metals that are mined in greater quantities. This distribution is made using the following formula for each element (Eq. 4.13):

$$Ton_i \ (\%) = \frac{m_i}{\sum_i^n m_i} \qquad (4.13)$$

where m_i is the tonnage of a specific metal with respect to the total tonnage of the set of elements that appear in the deposit.

Secondly, the allocation can be made using the market price of each metal in question. The sales price of metals tends to fluctuate over time, since it not only depends on extraction or demand but also political, socio-economic factors, etc. come into play. However, some studies affirm that the long-term price of metals remains more or less constant, although it may experience occasional variations (Gordon & Tilton, 2008). To create the cost distribution, we have used the annual average prices in dollars per ton updated to the last year taken and compiled by the USGS, ranging from 1900 to 2011 (Kelly & Matos, 2016). Thus, the price is calculated as follows (Eq. 4.14):

$$Price_i \ (\%) = \frac{\$_i \cdot m_i}{\sum_i^n \$_i \cdot m_i} \qquad (4.14)$$

where $\$_i$ represents the annual average price of the metal i.

Thirdly, the costs will be assigned using the exergy replacement costs (ERC), which vary depending on the metal considered. These costs determine the bonus that nature gives us by concentrating minerals in mines instead of having to extract them from common rocks in the Earth's crust. Let us remember that the greater the difference between the concentration in the crust and the concentration in the mines, and the greater the energy required to extract and process the metal, the greater the replacement costs. Therefore, they help to reflect the scarcity of each metal. To establish a comparison between the three factors analysed, we will need to multiply these replacement costs by the tonnage of each metal following this equation (Eq. 4.15):

$$ERC_i \ (\%) = \frac{ERC_i \cdot m_i}{\sum_i^n ERC_i \cdot m_i} \qquad (4.15)$$

Table 4.7 shows the values of the exergy replacement costs for each element studied in this case study, as well as the average prices for 1980 and 2006. We chose the average price data for 1980 since there was a global rebound in prices. Similarly, we also chose 2006, when similar events occurred.

Once we know how the cost distribution is going to be calculated in the three cases, we can study the different models of the most representative deposits. Specifically, the following types of deposits will be analysed: porphyry copper with gold, copper and nickel komatiites, massive Kuroko-type sulphide deposits and placer gold deposits and elements of the platinum group.

Let us begin the analysis with porphyry copper, which is the main source of copper worldwide. More than 60% of copper comes from these deposits. We will analyse the specific case of gold porphyry copper deposits, where the amount of gold present is higher than 30 g/t (Perello & Cabello, 1989). The best-known porphyry copper deposits with significant amounts of gold are in Papua New Guinea and Canada, although there are also other deposits in Argentina, Australia, Brazil, the USA and Indonesia.

According to studies by the United States Geological Survey, after analysing a total of 40 porphyry copper and gold deposits, they estimated that the average tonnage of these deposits was 100 million tons, with an average copper grade of 0.56% and an average gold concentration of 0.38 g per ton (USGS, 1986). In addition, small amounts of silver, between 1 and 4 g per ton, could appear along with molybdenum.

From this data, we can allocate costs for copper, gold and silver, since they are the majority elements in this type of porphyry deposit, by means of tonnage, price and exergy replacement costs. The results of the analysis are shown in Table 4.8. As we can see, copper is the element with the greatest weight if it is analysed in tonnage, exceeding 99.9%, which is logical when comparing the vast amounts of metal extracted compared to such small amounts of gold and silver. In price, in 1980

Table 4.7 Average prices of certain elements for 1980 and 2006 (USGS) and exergy replacement costs (ERC)

Element	Price 1980 ($/t)	Price 2006 ($/t)	ERC (GJ/t)
Cobalt	51,600	30,700	10,872
Copper	2,230	6,940	292
Nickel	6,230	24,200	524
Gold	19,700,000	19,500,000	553,250
Silver	663,300	373,000	7,371
Lead	937	1,710	37
Zinc	825	3,500	25

Table 4.8 Allocation of costs based on the tonnage, price and exergy replacement costs for porphyry copper (values in %) (Valero et al., 2015)

Element	Tonnage	Price		ERC
		1980	2006	
Copper	99.96	56.6	81.3	70.3
Gold	0.01	38	17.4	28.2
Silver	0.03	5.4	1.4	1.2

the costs were more divided between copper and gold, but in 2006, as the price of copper multiplied by three and the price of gold remained stable, copper again became more important by weight, which also strengthens the idea that the price is very volatile. However, in exergy replacement costs, a stable value over time, gold represents almost 30% in weight of the cost allocation since, although copper is mined in large quantities, the quality of the resource is much higher in the case of the precious metal.

Let us now consider the case of komatiites. After studying a total of 31 deposits, the United States Geological Survey deduced that the average tonnage for this type of deposit was 1.6 million tons, with an average nickel grade of 1.5% (USGS, 1986). Furthermore, the copper concentration in these deposits was between 0.09 and 0.26%, with other elements such as cobalt (with an average grade of 0.065%), gold (0.078 g per ton), palladium, platinum or iridium also appearing.

The allocation of costs according to the different methods explained is summarised in Table 4.9. In tonnage, nickel clearly predominates over the rest of the elements. In price, in 1980 nickel loses some weight compared to cobalt. However, due to the decrease in the price of cobalt per ton in 2006 and the increase in the sale price of nickel, nickel represents more than 90% in 2006. In ERC, although nickel continues to have a significant weight, cobalt, an element with an ERC almost 20 times higher than nickel, represents 36% of the cost allocation.

We can perform an equivalent exercise for massive sulphides, explained in Sect. 3.3. After studying a total of 432 massive sulphide deposits, the United States Geological Survey concluded that the average tonnage was 1.5 million tons, with an average concentration of 1.3% in copper (USGS, 1986). On the other hand, many

Table 4.9 Allocation of costs based on tonnage, price and exergy replacement costs for nickel and copper komatiites (values in %) (Valero et al., 2015)

Element	Tonnage	Price		ERC
		1980	2006	
Cobalt	3.5	24.0	4.7	35.9
Copper	8.2	2.4	2.5	0.8
Gold	0.0002	0.6	0.2	0.1
Nickel	88.3	73.0	92.7	63.2

other elements may appear in these deposits and, on average, it was found that they could contain 1.9% lead, between 2.0 and 6.7% zinc, between 0.16 and 2.3 g/t gold and between 13 and 100 g/t silver.

The allocation of costs according to the different methods explained is summarised in Table 4.10. In tonnage, the costs are divided between zinc, for the most part, followed by copper and lead, the three majority elements that are extracted. However, if the costs are distributed according to the price, certain variations are observed since the costs appear much more distributed among the five elements. In the case of the distribution in exergy replacement costs, the greatest weight corresponds to copper, followed by zinc and gold, a metal that is scarcer and more difficult to extract than, say, lead, which is almost in the last place.

Finally, let us study the case of placer-type gold deposits. Most of those studied in this section are of the quaternary age, that is, they were formed from 2.6 million years ago to the present day and are of alluvial origin, generating gold deposits from water currents. The United States Geological Survey analysed a total of 65 placer-type gold deposits distributed in different countries of the world, presenting an average tonnage of 1.1 million tons, a concentration of 0.2 g of gold per ton and 0.035 g of silver per ton (USGS, 1986).

As shown in Table 4.11, most of the cost allocation in tons corresponds to silver, but if we analyse by price, regardless of the year, the situation is reversed. In the case of exergy replacement costs, gold clearly prevails since the unit value of the exergy replacement cost is almost 75 times higher than that of silver.

From the analysed reservoir models, different conclusions can be drawn. First, in all cases, carrying out the cost allocation in tons does not help to obtain a true picture

Table 4.10 Allocation of costs based on tonnage, price and exergy replacement costs for massive sulphides (values in %) (Valero et al., 2015)

Element	Tonnage	Price		ERC
		1980	2006	
Copper	26.13	30.3	38.9	45.9
Gold	0.002	16.6	6.8	15.0
Lead	15.55	7.6	5.7	9.1
Silver	0.06	20.6	4.8	7.0
Zinc	58.26	25.0	43.8	23.0

Table 4.11 Allocation of costs based on tonnage, price and exergy replacement costs for PGE-gold placer deposits (values in %) (Valero et al., 2015)

Element	Tonnage	Price		ERC
		1980	2006	
Gold	13	81.7	88.7	92.2
Silver	87	18.3	11.3	7.8

of the effect that mineral depletion is having, since, although it is mathematically possible to add tons to grams, the quality aspect is ignored (see Fig. 4.9).

It has been possible to verify that the changes in prices mean that, depending on the moment in time that is considered, the greatest weight assigned is for one element or another. In some cases, these cost-sharing values do approximate those obtained with the exergy replacement costs; however, as shown in Fig. 4.10 for copper, the price variation can be very abrupt and variable, yet the exergy, and therefore exergy replacement costs, remains constant over time.

Therefore, if the exergy replacement costs are used to carry out this allocation of costs between the different elements that are extracted in a mine since it is a static value over time that takes into account not only the quantity but also the physical quality of the resource, the distribution is much fairer. Consequently, those scarce and difficult to extract minerals will have a higher cost compared to the more

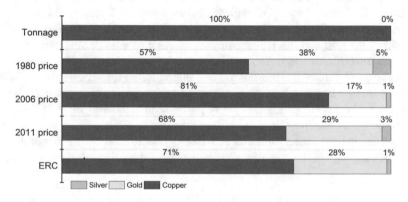

Fig. 4.9 Allocation of costs by different methods: tonnage, price and exergy replacement costs for a porphyry copper, gold and silver deposit (Valero et al., 2015)

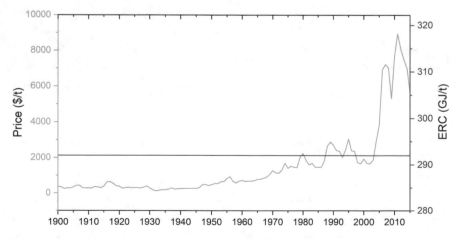

Fig. 4.10 Comparison between the price (in $/t) and ERC of copper (Valero et al., 2015)

abundant ones. In addition, the exergy replacement cost represents the last effort that future generations should take to extract the mineral when the deposits are depleted. This means that implicitly, future extraction costs are being considered, which could potentially be incorporated into the price of future raw materials. Ultimately, in the future, ERC-based allocation and price could converge, so a current ERC-based allocation somehow allows us to anticipate future market scenarios.

This allocation is also valid if it is carried out not only at the mine site but also in metallurgical processes. An exercise similar to that explained here was performed for the case of copper and nickel refining, with the purpose of verifying the method's validity (Valero et al., 2015). More possible applications of exergy replacement costs will be shown in future chapters.

4.8 Thermodynamic Rarity

So far we have developed and presented an application of exergy replacement costs through a grave-to-cradle approach. However, as explained above, it is just as important to consider the more traditional cradle-to-grave approach. The global vision of the life cycle of materials is achieved by uniting the two approaches and that is precisely what we intend through what we call *thermodynamic rarity*.

Generally speaking, when we talk about rarity, we tend to talk about a qualitative classification, for example, whether there is a small or large amount of a given element. But are we able to quantify it? The answer is yes, through exergy, with thermodynamic rarity.

The thermodynamic rarity (R) is the sum of two terms. First, the exergy replacement costs (ERC), that is, the total exergy needed to concentrate minerals from Thanatia to concentration in mines with the technology available today. The second term refers to the energy associated with the mining process, beneficiation and processing, and the exergy costs of the mine to the market (E_μ) (Fig. 4.11). That is, the thermodynamic rarity would be the total amount of exergy it takes to obtain a mineral from Thanatia to the beneficiation process, using the prevalent technologies available. Therefore, the rarer and more difficult to extract and refine a mineral, the greater its thermodynamic rarity. Similarly, a mineral that is easily processed but is very scarce in the crust, or vice versa, an abundant mineral but difficult to process, can also have high thermodynamic rarity. Thus, using the concept of thermodynamic rarity, the physical criticality of minerals can be analysed, depending on their quality, and can be measured in energy units such as kJ or ktoe.

The exergy replacement costs are actually avoided costs and they form the imaginary part of the mineral cycle. The real part is the exergy costs associated with the current extraction and beneficiation processes. For example, if the ore grade decreases in the mine (from x_{M1} to x_{M2}), the real extraction and beneficiation costs will increase. However, the exergy replacement costs (avoided costs) will decrease since the "bonus" that nature gives us will be reduced. Seen from a human perspective, at that moment it would be "easier" to replace resources to the initial state if

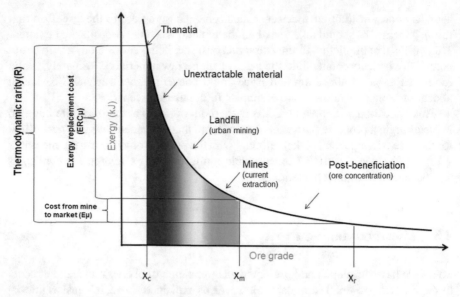

Fig. 4.11 Graphic definition of thermodynamic rarity

their original concentration is low (Fig. 4.12). This phenomenon is in fact what is expected in the future. Today, the grade of mineral deposits is high enough that it is more profitable to extract from nature than to recover materials from, for example, landfills. There will come a day when the laws and therefore the bonus will decrease

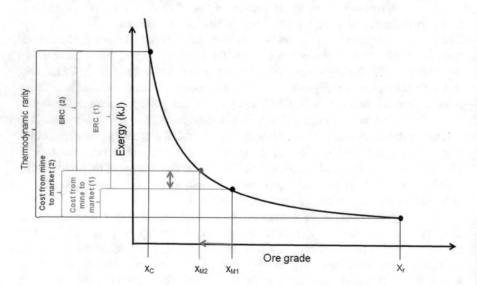

Fig. 4.12 Influence of extraction (or decrease in mining grade) on thermodynamic rarity

enough so that the extraction in landfills (known as urban mining) has costs comparable to those of traditional mining (the central yellow-orange stripe in Fig. 4.11). At the limit, when all the minerals have been dispersed and the deposits depleted, the avoided and imaginary costs represented through the ERC will be converted completely into real costs. We will have completely exhausted the mineral bonus and humans will be forced to extract them from the common rocks themselves, that is, from Thanatia.

It should be remembered that the thermodynamic rarity of a mineral is a function of its concentration in nature and technology. As a general rule, if the technology does not change, the thermodynamic rarity of a particular mineral will remain constant over time, since it only depends on the initial and final state of Thanatia and of the quality of the resource after processing—typically a value set by the market. If there are improvements in the available extraction and processing technologies, the thermodynamic rarity will decrease, due to the decrease in the exergy costs of the extraction and processing and the exergy replacement costs (since equivalent technologies are used).

To calculate the thermodynamic rarity of a mineral, we need to know the concentration of the mineral in Thanatia (x_c), the average concentration in the mines (x_m) and the amount of energy used in the extraction process (E_{min}) and beneficiation (E_{ben}) of ore with current technologies based on ore grade. The concentration data is known, so the missing values are those related to the energy required during the extraction and processing of the mineral. An exhaustive description of what each recovery process consists of for the different elements and their corresponding energy can be found in (Valero & Valero, 2014). Given that the concept of thermodynamic rarity was created with the objective of being an indicator of scarcity, and given also that if technological improvements occur, they will be applied progressively to all minerals, we can maintain the calculated values assuming the technologies currently applicable. This does not mean that they can be updated in the future if better information is available or if new technologies emerge that appreciably change values.

Next, by way of example, we will calculate the thermodynamic rarity of cadmium. The value of the replacement cost of cadmium calculated with the procedure explained in the previous section is 5,898 GJ/t, so the only thing left to know is the energy consumption of the extraction, processing and beneficiation. As previously mentioned, this type of information is difficult to obtain, but there are some studies and reports in which average values appear (Botero, 2000; Chapman & Roberts, 1983; IPPC, 2002). In the case of cadmium, the energy associated with extraction and concentration is 264 GJ/t and the energy associated with the beneficiation process is 279 GJ/t (IPPC, 2002). Therefore, the thermodynamic rarity will be the sum of these three values, that is, 6,441 GJ/t. Table 4.12 shows a summary of all this data for the different minerals.

In Fig. 4.13, those elements of the periodic table that have higher values of thermodynamic rarity can be seen with a colour code. Red represents those elements that have high thermodynamic rarity values, which are cobalt, palladium, platinum, gallium, germanium and tellurium, among others. These elements share the fact that they are scarce in nature and that the energy required to obtain them is high. At the

Table 4.12 Concentration in Thanatia (x_c), average concentration in current mines (x_m), energy associated with the extraction and concentration process (E_{min}), energy associated with the beneficiation process (E_{ben}), exergy replacement costs (ERC) and thermodynamic rarity (R) (Valero & Valero, 2014)

Element	Ore	x_c	x_m	E_{min}	E_{ben}	ERC	R
		g/g	g/g	GJ/t	GJ/t	GJ/t	GJ/t
Aluminium (bauxite)	Gibbsite	1.38×10^{-3}	7.03×10^{-1}	31	24	627	682
Antimony	Stibnite	2.75×10^{-7}	5.27×10^{-2}	1.4	12	474	488
Arsenic	Arsenopyrite	4.71×10^{-6}	2.17×10^{-2}	9	19	400	428
Barium	Baryte	7.09×10^{-4}	9.50×10^{-1}	0.9	–	38	39
Beryllium	Beryl	3.22×10^{-5}	7.80×10^{-2}	7	450	253	710
Bismuth	Bismuthinite	5.10×10^{-8}	2.46×10^{-3}	4	53	489	546
Cadmium	Greenockite	1.16×10^{-7}	1.28×10^{-4}	264	279	5,898	6,441
Limestone	Calcite	8.00×10^{-3}	6.00×10^{-1}	0.4	6	3	9
Cerium	Monazite	1.03×10^{-4}	3.00×10^{-4}	523	–	97	620
Zirconium	Zircon	3.88×10^{-4}	4.02×10^{-3}	739	633	654	2,026
Cobalt	Linnaeite	5.15×10^{-9}	1.90×10^{-3}	9	129	10,872	11,010
Copper	Chalcopyrite	6.64×10^{-5}	1.67×10^{-2}	35	21	292	348
Chromium	Chromite	1.98×10^{-4}	6.37×10^{-1}	0,1	36	5	41
Platinum group elements	Cooperite	3.95×10^{-10}	8.02×10^{-7}	175,000	–	2,695,013	2,870,013
Fluorspar	Fluorite	1.12×10^{-4}	2.50×10^{-1}	1,5	–	183	184
Tin	Cassiterite	2.61×10^{-6}	6.09×10^{-3}	15	11	426	453
Strontium	Celestite	6.70×10^{-4}	4.00×10^{-1}	0,2	72	4	76
Gadolinium	Monazite	1.30×10^{-4}	3.00×10^{-4}	3607	–	478	4,085
Gallium	Ga in bauxite	1.76×10^{-5}	5.00×10^{-5}	610,000	–	144,828	754,828

(continued)

Table 4.12 (continued)

Element	Ore	x_c g/g	x_m g/g	E_{min} GJ/t	E_{ben} GJ/t	ERC GJ/t	R GJ/t
Germanium	Ge in zinc	1.41×10^{-6}	3.00×10^{-3}	498	–	23,749	24,247
Graphite	Graphite	2.41×10^{-4}	1.50×10^{-1}	1.1	–	20	22
Iron	Hematite	9.66×10^{-4}	7.30×10^{-1}	0.7	13	18	32
Indium	In (in zinc)	5.61×10^{-8}	4.50×10^{-4}	3,320	–	360,598	363,917
Yttrium	Monazite	1.30×10^{-4}	3.00×10^{-4}	1,198	–	159	1,357
Lanthanum	Monazite	1.30×10^{-4}	3.00×10^{-4}	297	–	39	336
Lithium	Spodumene	3.83×10^{-4}	8.04×10^{-1}	13	420	546	978
Magnesium	Magnesite	1.28×10^{-3}	4.20×10^{-1}	10	437	136	583
Manganese	Pyrolusite	4.90×10^{-5}	5.00×10^{-1}	0.2	57	16	73
Mercury	Cinnabar	5.73×10^{-8}	4.41×10^{-3}	157	252	28,298	28,707
Molybdenum	Molybdenite	1.83×10^{-6}	5.01×10^{-4}	136	12	908	1,056
Neodymium	Monazite	1.30×10^{-4}	3.00×10^{-4}	592	–	78	670
Niobium	Ferrocolumbite	8.10×10^{-6}	2.00×10^{-2}	132	228	4,422	4,782
Nickel	Pentlandite	5.75×10^{-5}	3.36×10^{-2}	15	100	761	877
Nickel	Garnierite	4.10×10^{-6}	4.42×10^{-2}	1.7	412	167	581
Gold	Native gold	1.28×10^{-9}	2.24×10^{-6}	110,057	–	553,250	663,307
Silver	Argentite	1.24×10^{-8}	4.27×10^{-6}	1,281	285	7,371	8,938
Lead	Galena	6.67×10^{-6}	2.37×10^{-2}	0.9	3	37	41

(continued)

Table 4.12 (continued)

Element	Ore	x_c	x_m	E_{min}	E_{ben}	ERC	R
		g/g	g/g	GJ/t	GJ/t	GJ/t	GJ/t
Potassium	Sylvite	2.05×10^{-6}	3.99×10^{-1}	1.7	–	665	667
Praseodymium	Monazite	7.10×10^{-6}	3.00×10^{-4}	296	–	577	873
Phosphate rock	Fluorapatite	4.03×10^{-4}	5.97×10^{-3}	0.3	5	0.35	5
Silica	Quartz	2.29×10^{-1}	6.50×10^{-1}	0.7	76	0.7	77
Sodium	Salt	5.89×10^{-4}	2.00×10^{-1}	1.3	40	17	58
Tantalum	Tantalite	1.58×10^{-7}	7.44×10^{-3}	3,083	8	482,828	485,919
Tellurium	Tetradymite	5.00×10^{-9}	1.00×10^{-6}	589,366	39	2,235,699	2,825,104
Rare Earths	Bastnaesite	2.54×10^{-7}	6.00×10^{-2}	10	374	348	733
Titanium	Ilmenite	4.71×10^{-3}	2.42×10^{-2}	7	128	5	140
Titanium	Rutile	2.73×10^{-4}	2.01×10^{-3}	14	244	9	266
Tungsten	Scheelite	2.67×10^{-6}	8.94×10^{-3}	213	144	7,429	8,023
Vanadium	V in different ores	9.70×10^{-5}	2.00×10^{-2}	136	381	1,055	1,572
Gypsum	Gypsum	1.26×10^{-4}	8.00×10^{-1}	0.2	–	15	16
Zinc	Sphalerite	9.96×10^{-5}	6.05×10^{-2}	1.5	40	25	67

other extreme, in green, are those that have values below 100 GJ/t, such as calcium, barium, iron or silicon, abundant and easy to obtain elements.

In summary, thermodynamic rarity can be considered as an indicator of the physical criticality of the elements. The expression grave to cradle, that is to say, the exergy replacement cost implies the idea of keeping safe which is scarce, while cradle to grave implies the idea of improving the efficiency of the extraction and refining processes. The greater the thermodynamic rarity of a certain element, the more reasons we will have to manage it properly, seeking alternatives to replace it with more abundant elements, avoiding its dispersion and optimising the processes for obtaining said element as they are very energy-intensive.

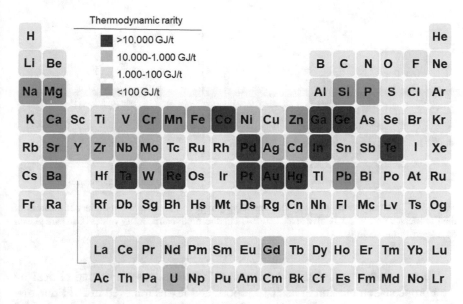

Fig. 4.13 Periodic table where the thermodynamic rarity of the elements has been represented with a colour code. *Source* Stanek et al. (2017)

4.8.1 Thermodynamic Rarity as an Indicator of Criticality

As mentioned above, the concept of thermodynamic rarity arises with the objective of being an indicator of scarcity. As such, we can compare it with other indicators of criticality, such as those established by the European Commission in its reports explained in Chap. 2 (European Commission, 2014, 2017, 2020). Neither the supply risk nor the economic importance considers an aspect as relevant as the physical reality of the resource, which can be measured through thermodynamic rarity. Through this comparison, a triple-axis (rarity, economic importance and supply risk) can be established so that a material can be considered critical if it is critical in at least two of the three axes.

The thresholds for determining the criticality of an element of the European Commission for the 2014 report was 1.0 in the case of supply risk and 5 in the case of economic importance. For thermodynamic rarity, a value of $1,000 \pm 5\%$ GJ/t has been considered, so that, if the value of an element is higher, it can be considered to be critical (Calvo et al., 2017b).

When confronting thermodynamic rarity with supply risk (Fig. 4.14), it is found that some of the minerals analysed, such as gold, tantalum or tellurium, have high values of thermodynamic rarity, since they are scarce minerals and difficult to extract. A similar situation can be observed when comparing thermodynamic rarity with economic importance. In the case of gold, although rare from a thermodynamic point of view, it is not critical in terms of economic importance nor in terms of risk of supply, so it does not meet the established criteria.

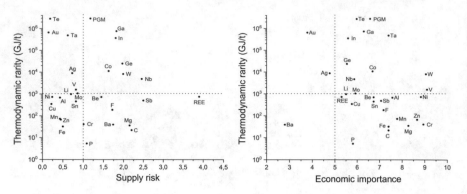

Fig. 4.14 Thermodynamic rarity against the supply risk and economic importance indicators of the European Commission (Calvo et al., 2017b). The thermodynamic rarity values (GJ/t) are placed on a logarithmic scale

We also analyse tantalum and tellurium. Tantalum was considered critical by the European Commission in its 2010 report, and it was removed in 2014 due to a decrease in the risk of supply, and in the 2017 and 2020 report, it was again included as critical. In the case of tellurium, it has never been included in the reports as critical. It is also striking that some substances, such as graphite or chromium, which have low thermodynamic rarity values, are considered critical. For example, graphite is considered critical since it is used in many sectors and because it cannot be substituted in many of these applications by other elements.

Therefore, by combining thermodynamic rarity with the European Commission's methodology to develop a list of critical materials, a list can be obtained that also takes into account the scarcity and quality of resources. The list of elements that are considered critical in at least two of the three axes considered is shown in Table 4.13. The elements marked with an asterisk are those not considered critical by the European Commission in its 2014 report but would be critical if the thermodynamic rarity were taken into account.

If these results of Table 4.13, based on the 2014 report, are compared with the last report published in 2020 by the European Commission, we can see that chromium was

Table 4.13 List of elements considered critical combining thermodynamic rarity with supply risk and economic importance of the European Commission (2014 report)

Antimony	Fluorspar	Lithium	Tantalum
Beryllium	Gallium	Magnesite	Tellurium
Cobalt	Germanium	Molybdenum	Rare Earth elements
Chromium	Graphite	Niobium	Tungsten
Platinum group elements	Indium	Phosphate rock	Vanadium

no longer considered critical, neither was graphite or magnesite (European Commission, 2020). On the other hand, tantalum was indeed included in the 2020 criticality list, as well as lithium.

In short, thermodynamics allows us to establish criticalities in the mid and long term in a predictable way, as has happened in this case. This is so because the criticalities sooner or later end up converging with what the physical reality of the resource dictates since it is the physical limits that are quite likely to finally prevail.

4.8.2 The Need for "Recyclaiming"

The thermodynamic rarity and the exergy replacement costs can also be useful to properly assess the raw materials found in the waste, as we will see below. At the beginning of the cycle of any element, the natural stock is extracted at the processing stage, which includes mining and metal refining, generating waste in the form of dumps. The refined metal goes to the manufacturing process, in which a series of waste is also generated, and which ends with the product in use. The recycling stage, in turn, is made up of different processes that involve both logistics and physicochemical processes (Fig. 4.15).

All these stages can be analysed as if they were a series of chemical reactions, with their corresponding exergy exchanges (Valero & Valero, 2019a). In the first place,

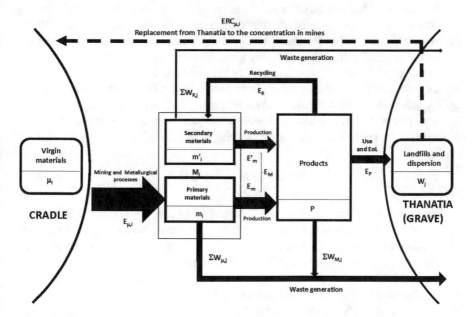

Fig. 4.15 Exergy processes and costs involved in the production and end of life of a product (Valero & Valero, 2019a)

the manufacturing process of an object or product can be considered as a chemical reaction that requires a manufacturing energy cost (or exergy) (E) that will depend on the materials (M_1, M_2, M_3...) that make up the product (called P) and that, in turn, inevitably generate waste (called $W_{M,j}$) (Eq. 4.16).

$$M_1 + M_2 + M_3 + \cdots + E_M \rightarrow P + \sum W_{M,j} \qquad (4.16)$$

These materials can themselves be of primary origin (m_i) or secondary origin (m'_i); each type of material, primary or secondary, will in turn be associated with different processing energy (E_m; E'_m) whose sum will constitute the total energy cost (E_M) of manufacturing.

The recycling process involves the disintegration of the product, which can also be represented as follows (Eq. 4.17):

$$P + E_R \rightarrow m'_1 + m'_2 + m'_3 \ldots + \sum W_{R,j} \qquad (4.17)$$

E_R would be the exergy cost of recycling, measured in kWh, which would mean separating and extracting all the materials that compose it through different processes of disassembly, melting down, etc.

If the result of the recycling process obtains the same quality as the original materials, then the quantity of secondary material would be equal to that of primary material, that is, $m'_1 = m_i$. In practice, this is not usually the case and we talk of downcycling, which means recycling something in such a way that the resulting product has a lower value than the original article. The recycling process will also generate a series of residues since it's never the case that the quantity of material recovered is equal to the initial quantity used to manufacture the product.

So the question now is, which is better? Creating a new product P from recycled materials or from materials extracted from nature? For this, we can use the recoverability index (RC), which measures the relationship between $E\mu$, the sum of exergies necessary to extract from the natural environment all the primary materials that have been part of the product (P), and the exergy cost of recycling (E_R), following Eq. 4.18:

$$RC_{M,total} = \frac{E_\mu}{E_R} * 100 \qquad (4.18)$$

For example, an 80% recoverability would mean that we are very close to full recovery. A ratio above 100%, would be ideal and should be the long-term goal of all objects, that is, their ability to be recovered to create new objects. Unfortunately, this figure rarely reaches very high levels in practice.

Recycling efforts, except for traditional metals such as iron, copper or aluminium, have focused on avoiding contamination, rather than on recovering the valuable materials they contain. In fact, the metal with the highest recycling rate today is lead, which is found mainly in car batteries and for which there is very strict recycling legislation. Fluorescent tubes must also undergo a recycling process, not to obtain the valuable elements they contain such as rare earth elements and phosphorous that are not recycled, but to prevent the mercury they contain from contaminating the environment. The result is that minor but valuable metals end up in landfills or downcycled with other metals such as steel and therefore lose their original functionality.

Recovery processes should be all the more critical the rarer the alloyed metals are. The scarcity of these metals can be assessed through the thermodynamic rarity of each metal (R_μ) which is made up of the exergy replacement cost (ERC_μ) plus extraction and refining costs E_μ (Sect. 4.8). Therefore, one way of considering the "need for recycling" or "recyclaiming" would be to compare it with its rarity rather than with its energy cost of extraction and processing (Valero & Valero, 2019a, 2019b). The greater the rarity of the component in question and the more energetically simple its recovery, the more interest there will be for it to be recycled. An example could be gold from vehicle electronics or smartphones. The expression needed to calculate the recyclability (RC_R) would be (Eqs. 4.19 and 4.20):

$$RC_R = \left[1 - \frac{E_R}{R_\mu}\right] * 100 \qquad (4.19)$$

where

$$R_\mu = R_{\mu 1} + R_{\mu 2} + R_{\mu 3} + \cdots = ERC_\mu + E_\mu \qquad (4.20)$$

The recycling process will be favoured if $E_R < R_\mu$. As the rarity is always greater than the extraction, beneficiation and refining costs, recycling will be much more beneficial.

In short, the need for recycling material will be greater, the greater its rarity (R_μ). The circular economy, which we will see in more detail in Chap. 8, would provide a product with many more lives at the end of its life if rarity were used as a criterion of need for recycling, the objective would then be to recover the rarest metals and not the most profitable. Consequently, the products could be eco-designed to recover the maximum of valuable materials.

Unfortunately, neither companies nor policymakers think about the avoided costs that nature provides us with for free. Present generations dismiss these costs as imaginary, however, future ones will see them as real as the mining grades gradually decrease. This would be resolved with an environmental fiscal policy that covers all, or at least in part, the avoided costs. In fact, a long-term recycling policy cannot be based on market fluctuations and oscillations, since the prices of raw materials do not adequately reflect the depletion of mineral deposits (Valero & Valero, 2019a).

Now that we know how to evaluate the loss of mineral wealth based on thermodynamics, through exergy replacement costs and thermodynamic rarity, the time has come to put it into practice—we will see this in the next chapter.

References

Ahrendts, J. (1977). *The exergy of chemically reacting systems*. Dusseldorf.

Ahrendts, J. (1980). Reference states. *Energy, 5*, 667–677.

Bosjankovic, F. (1963). Reference level of exergy of chemically reacting systems. *Forschung Im Ingenieurwesen, 21*, 151–152.

Botero, E. (2000). *Valoración exergética de recursos naturales, minerales, agua y combustibles fósiles*. Ph.D. thesis, Universidad de Zaragoza.

BP. (2019). British petroleum. *Statistical Review of World Energy*. Available at: https://www.bp.com/content/dam/bp/business-sites/en/global/corporate/pdfs/energy-economics/statistical-review/bp-stats-review-2019-full-report.pdf.

Calvo, G., Mudd, G., Valero, A., & Valero, A. (2016). Decreasing ore grades in global metallic mining: A theoretical issue or a global reality? *Resources, 5*(4).

Calvo, G., Valero, A., & Valero, A. (2017a). Assessing maximum production peak and resource availability of non-fuel mineral resources: Analyzing the influence of extractable global resources. *Resources, Conservation and Recycling, 125*, 208–217.

Calvo, G., Valero, A., & Valero, A. (2017b). Thermodynamic approach to evaluate the criticality of raw materials and its application through a material flow analysis in Europe. *Journal of Industrial Ecology*. https://doi.org/10.1111/jiec.12624.

Chapman, P. F. & Roberts, F. (1983). *Metal resources and energy*. Butterworths.

Clarke, F. W. (1889). The relative abundance of the chemical elements. *Bulletin of the Philosophical Society of Washington, XI*, 131–142.

Condie, K. C. (1993). Chemical composition and evolution of the upper continental crust: Contrasting results from surface samples and shales. *Chemical Geology, 104*, 1–37.

Diederichsen, D. (1999). *Referenzumgebungen zur Berechnung der chemischen Exergie, Tech. rep., Technical Report, Fortschr.-Ber.VDI Reihe 19, Düsseldorf, in German*.

El-Sayed, Y., & Aplenc, A. (1970). Application of the thermoeconomic approach to the analysis and optimization of a vapor-compression desalting system. *Journal of Engineering for Power,* 17–26.

European Commission. (2014). *Report on critical raw materials for the EU. Report of the Ad hoc working group on defining critical raw materials*.

European Commission. (2017). *Study on the review of the list of Critical Raw Materials. Critical raw materials factsheets*.

European Commission. (2020). *Study on the EU's list of Critical Raw Materials (2020). Final Report*.

Evans, R. B., & Tribus, M. (1965). Thermo-economics of saline water conversion. *Industrial & Engineering Chemistry Process Design and Development, 4*, 195–206.

Gaggioli, R. A., & Petit, P. J. (1976). Second Law analysis for pinpointing the true inefficiencies in final conversion systems. *A.C.S. Division of Fuel Chemistry, 21*(2).

Gordon, R. L., & Tilton, J. E. (2008). Mineral economics: Overview of a discipline. *Resources Policy, 33*, 4–11.

Grigor'ev, N. A. (2007). Average composition of the upper continental crust and dimensions of the maximum concentration of chemical elements. {Geology of the Ural and neighbouring territories. Summary materials 2002–2006}. *Uralian Geological Journey*.

IPPC. (2002). *Integrated Pollution Prevention and Control (IPPC); Draft reference document on best available techniques for management of tailings and waste-rock in mining activities*.

Kelly, T. D., & Matos, G. R. (2016). *Historical statistics for mineral and material commodities in the United States: U.S. Geological Survey Data Series 140*. Available at: https://www.usgs.gov/centers/nmic/historical-statistics-mineral-and-material-commodities-united-states.

Kennecott Utah Copper Corporation. (2004). *Copper environmental profile. Life cycle assessment*.

Lovelock, J. E. (1972). Gaia as seen through the atmosphere. *Atmospheric Environment, 6*(8), 579–580.

Lovelock, J. (2006). *The revenge of Gaia: Why the Earth is fighting back—And how we can still save humanity*. Penguin Press.

Mudd, G. M. (2007). Global trends in gold mining: Towards quantifying environmental and resource sustainability. *Resources Policy, 32*(1–2), 42–56.

Mudd, G. M. (2010). Global trends and environmental issues in nickel mining: Sulfides versus laterites. *Ore Geology Reviews, 38*(1–2), 9–26.

Norgate, T., & Haque, N. (2012). Using life cycle assessment to evaluate some environmental impacts of gold production. *Journal of Cleaner Production, 29–30*, 53–63.

Palacios, J. L., Abadías, A., Valero, A., Valero, A., & Reuter, M. A. (2019). Simulation-based approach to study the effect of the ore-grade decline on the production of gold. In *ECOS 2019—The 32th International Conference on Efficiency, Cost, Optimization, Simulation and Environmental Impact of Energy Systems*, Wroclaw, Poland.

Perello, J., & Cabello, J. (1989). Pórfidos cupríferos ricos en oro: una revisión. *Revista Geológica De Chile, 16*(1), 73–92.

Ranz, L. (1999). *Análisis de los costes exergéticos de la riqueza mineral terrestre. Su aplicación para la gestión de la sostenibilidad*. Ph.D. thesis, Universidad de Zaragoza.

Ronov, A. B., & Yaroshevsky, A. A. (1969). *Earth's crust and upper mantle* (vol. 13). Washington D.C.: American Geophysical Union.

Rudnick, R. L., & Gao, S. (2004). Composition of the continental crust. In R. L. Rudnick (Ed.), *Treatise on geochemistry. The crust* (vol. 3, pp. 1–64). Elsevier Pergamon.

Stanek, W., Valero, A., Valero, A., Uche, J., & Calvo, G. (2017). *Thermodynamic methods to evaluate resources*. In: Stanek W. (eds) Thermodynamics for Sustainable Management of Natural Resources. Green Energy and Technology. Springer, Cham. https://doi.org/10.1007/978-3-319-48649-9_6.

Sussman, M. V. (1979). Choosing a reference environment-state for available-energy computations. In *72nd Annual AIChE Meeting, San Francisco, California, 11/79*.

Szargut, J. (1987). Standard chemical exergy of some elements and their compounds, based upon the concentration in Earth's crust. *Geochemistry International, 35*(1–2), 53–60.

Szargut, J. (1989). Chemical exergies of the elements. *Applied Energy, 32*, 269–286.

Tavares, L. M., & King, R. P. (1998). Single-particle fracture under impact loading. *International Journal of Mineral Processing, 54*(1), 1–28.

Tribus, M., & Evans, R. (1962). *The thermoeconomics of sea water conversion*. UCLA Report No. 62–63.

Tromans, D., & Meech, J. A. (2002). Fracture toughness and surface energies of minerals: Theoretical estimates for oxides, sulphides, silicates and halides. *Minerals Engineering, 15*(12), 1027–1041.

USGS. (1986). *Mineral deposit models. Bulletin 1693. United States Geological Survey*. In: D. P. Cox & D. A. Singer (Eds.) (vol. 379).

USGS. (2019). *Mineral commodity summaries 2019. United States Geological Service*. Available at: https://prd-wret.s3-us-west-2.amazonaws.com/assets/palladium/production/atoms/files/mcs2019_all.pdf.

Valero, A. (1998). *Thermoeconomics as a conceptual basis for energy-ecological analysis*. In S. Ulgiati (Ed.), *Advances in energy studies. Energy flows in ecology and economy* (pp. 415–444), Musis, Roma.

Valero, A. (2008). *Exergy evolution of the mineral capital on Earth*. Ph.D. thesis, University of Zaragoza, Zaragoza, Spain.

Valero, A., Agudelo, A., & Valero, A. (2011a). The crepuscular planet. A model for the exhausted atmosphere and hydrosphere. *Energy, 36*(6), 3745–3753.

Valero, A., Dominguez, A., & Valero, A. (2015). Exergy cost allocation of by-products in the mining and metallurgical industry. *Resources, Conservation and Recycling, 102*, 128–142.

Valero, A., Lozano, M., & Muñoz, M. (1986). *A general theory of exergy saving. I. On the exergetic cost*. In: R. Gaggioli (Ed.), *Computer-aided engineering and energy systems. Second law analysis and modelling* (vol. 3, pp. 1–8).

Valero, A., & Valero, A. (2010). Exergoecology: A thermodynamic approach for accounting the Earth's mineral capital. The case of bauxite-aluminium and limestone-line chains. *Energy, 1*, 229–238.

Valero, A., & Valero, A. (2012a). What are the clean reserves of fossil fuels? *Resources, Conservation and Recycling, 68*, 126–131.

Valero, A., & Valero, A. (2012b). Exergy of comminution and the Thanatia Earth's model. *Energy, 44*(1), 1085–1093.

Valero, A., & Valero, A. (2014). *Thanatia: The destiny of the Earth's mineral resources: A thermodynamic cradle-to-cradle assessment.* World Scientific Publishing Company.

Valero, A., & Valero, A. (2019a). Thermodynamic rarity and recyclability of raw materials in the energy transition: The need for an in-spiral economy. *Entropy, 21*(9), 873.

Valero, A., & Valero, A. (2019b). Pensando más allá del primer ciclo: economía espiral. In *Economía Circular-Espiral: Transición hacia un metabolismo económico cerrado* (p. 343). Ecobook.

Valero, A., Valero, A., & Gómez, B. J. (2011b). The crepuscular planet. A model for the exhausted continental crust. *Energy, 36*(6), 694–707.

Valero, A., Valero, A., & Domínguez, A. (2013). Exergy replacement cost of mineral resources. *Journal of Environmental Accounting and Management, 1*(1), 147–158.

van Gool, W. (1998). Thermodynamics of chemical references for exergy analysis. *Energy Conversion and Management, 39*(16–18), 1719–1728.

Wall, G. (1977). *Exergy—A useful concept within resource accounting.* Institute of Theoretical Physics, Göteborg, Report(77-42).

Ward, P. (2009). *The Medea hypothesis: Is life on earth ultimately self-destructive?* Princeton University Press.

Chapter 5
Thermodynamic Assessment of the Loss of Mineral Wealth

Abstract In this chapter, we will analyse the global historical extraction of mineral resources from an exergy point of view. Exergy allows us to assess the degradation caused by the extraction of minerals by humans considering the resource quality. The loss of total mineral wealth from 1900 to 2018 was around 200 Gtoe. This means that we would need a minimum of 80% of the world oil reserves known today to replace or recover those mineral deposits from a dispersed environment (in our case, Thanatia) with current technology. Additionally, the future situation of mineral extraction can be contemplated using projections based on the Hubbert model, using past trends and resource estimates. The case of lithium and phosphate rock will be explored in detail. Finally, using Sankey diagrams, mineral reliance and loss of mineral wealth for three regions will be analysed: (1) Spain and Colombia, (2) Latin America and (3) Europe. This, combined with an economic analysis, will help us to detect if these countries' natural heritage is being undersold.

In the previous chapters, we have seen what the current situation of the demand for minerals is. Additionally, we have shown that exergy is a very useful tool for evaluating the loss of mineral wealth of a country, region or continent using the same unit for both fuel and non-fuel mineral resources. It allows us to concretely assess the degradation caused by the extraction of minerals by humans throughout the last decades and even centuries. Unlike the mass indicator, the most common tool used in this type of calculation, all the characteristics that describe a mineral resource can be accounted for in a unified manner through the use of exergy: the amount extracted, its chemical composition, and its concentration (ore grade). The exergy replacement costs also provide numbers closer to the social perception of the value of minerals. We are thus able to put a figure (strictly based on physical aspects) to the amount of energy that nature provides us by concentrating minerals in deposits and thus accounting for the impact of their depletion.

We will now demonstrate how the global historical extraction of the various minerals can be evaluated from an exergy point of view, thus considering the resource quality. We will also look at how to estimate future extraction based on Hubbert's model. Finally, we will analyse the mineral exergy balance of different regions to assess the degree of sustainability associated with resource extraction, import and export.

5.1   Exergy Evolution of Global Historical Mineral Extraction

To assess the degradation of the planet's mineral wealth, we first need to collect data on the primary extraction of the different minerals worldwide. This is not a simple task since the extraction of certain specific metals at a global level is unknown. This is particularly true for by-products from the metallurgical industry, such as tellurium or selenium, for which only estimates from the United States Geological Survey (USGS) or the British Geological Survey (BGS) exist. In other cases, such as rare earths, global extraction figures are not entirely reliable since illegal extraction in China is high and is not usually included in the statistics due to the many associated unknowns. However, in terms of the most abundant mineral resources—aluminium, iron or common salt—the data reflect reality more adequately.

Furthermore, these historical global statistics of mineral extraction usually date back only to the beginning of the twentieth century. However, there are mining statistics from certain countries that go as far back as the mid-nineteenth century. The Spanish Mining Statistics, for example, has been published uninterruptedly since 1861. Even so, as not enough information exists to be able to apply insights globally, the period studied in the case study below ranges from 1900 to the most recent year available.

Figures 5.1 and 5.2 show the extraction of the vast majority of mineral resources in this period of time, expressed in Mtoe (in exergy replacement costs); data have been separated into two figures so that those substances that are produced in smaller quantities by the massive extraction of aluminium or iron are not concealed.

When expressing the data in this unit, although iron ore is the world's most extracted mineral by weight, it is aluminium that occupies the first position when resource quality is considered. Also interesting in Fig. 5.2 is the fact that platinum group elements (PGM) are especially relevant, since their mass extraction is of the order of a few hundred tons. Fluorspar (also known as fluorite), in this same figure, has a weight like that of PGM and yet its mass extraction is around six million tons. Since PGMs are much scarcer minerals and require much more processing than fluorspar, it is logical that it acquires a much more relevant value when expressed in exergy terms and that the associated mineral wealth loss is much more significant.

Thus, taking into account the extraction data, we can also estimate the loss of mineral wealth (LMW) or the mineral depletion caused by the extraction of the last 118 years (Table 5.1). The minerals with the highest accumulated extraction in this period are iron, salt, limestone, aluminium and gypsum, exceeding several billion tons. Also noteworthy are gallium, germanium and indium values, which did not begin to be extracted in a significant way until the second half of the twentieth

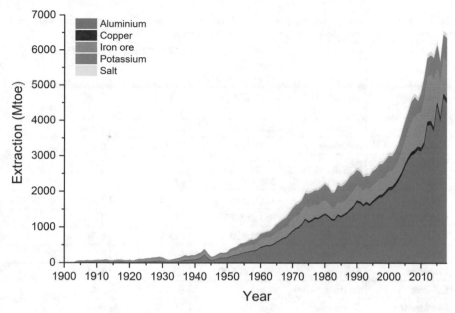

Fig. 5.1 Extraction of five elements and minerals expressed in Mtoe. Data source: USGS

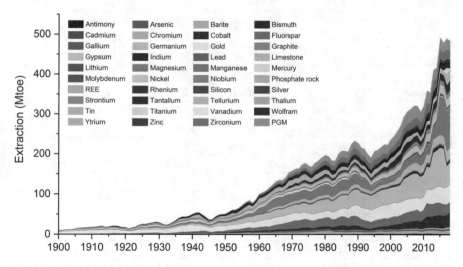

Fig. 5.2 Extraction of the rest of the elements and minerals expressed in Mtoe. Data source: USGS

century. Their accumulated extraction already reaches thousands of tons despite being always used in very small quantities in products.

If we look at the data expressed in the loss of mineral wealth, that is, using exergy replacement costs, the minerals with the highest values are aluminium, iron,

Table 5.1 Global depletion of the main mineral resources, expressed in exergy terms and sorted in decreasing order according to the loss of total mineral wealth (LMW)

Element	Ore	Cumulative extraction (1900–2018) (t)	LMW (Mtep)	LMW (Mtep/year)
Aluminium (bauxite)	Gibbsite	7,494,299,000	111,940.5	940.7
Antimony	Stibnite	7,513,190	84.9	0.7
Arsenic	Arsenopyrite	4,091,930	39.0	0.3
Barium	Baryte	392,831,000	358.6	3.0
Bismuth	Bismuthinite	346,857	4.0	<0.1
Cadmium	Greenockite	1,253,386	176.0	1.5
Chromium	Chromite	857,963,700	92.7	0.8
Cobalt	Linnaeite	2,996,430	775.6	6.5
Copper	Chalcopyrite	707,998,000	4,917.2	41.3
Fluorspar	Fluorite	303,185,500	1,318.5	11.1
Gallium	Ga (in bauxite)	4,247	14.6	0.1
Germanium	Ge (in zinc)	5,089	2.9	<0.1
Gold	Native gold	162,163	2,136.1	18.0
Graphite	Graphite	51,714,600	25.1	0.2
Gypsum	Gypsum	6,676,980,000	2,449.7	20.6
Indium	In (in zinc)	13,908	119.4	1.0
Iron	Hematite	83,455,900,000	35,271.7	296.4
Lead	Galena	265,645,000	231.6	1.9
Limestone	Calcite	9,696,000,000	603.9	5.1
Lithium	Spodumene	644,9	8.4	0.1
Magnesium	Magnesite	458,528,395	1,487.0	12.5
Manganese	Pyrolusite	682,629,000	254.2	2.1
Mercury	Cinnabar	573,57	386.4	3.2
Molybdenum	Molybdenite	8,221,232	177.7	1.5
Nickel	Pentlandite and Garnierite	68,827,690	663.5	5.6
Niobium	Ferrocolumbite	1,447,000	152.3	1.3
Phosphate rock	Fluorapatite	8,934,500,000	75.0	0.6
Platinum group elements	Cooperite	17,197	1,103.5	9.3
Potassium	Sylvite	1,654,850,000	26,198.7	220.2
Rare earth elements	Bastnaesite	3,748,921	31.1	0.3
Silicon	Quartz	204,750,000	3.6	<0.1

(continued)

Table 5.1   (continued)

Element	Ore	Cumulative extraction (1900–2018) (t)	LMW (Mtep)	LMW (Mtep/year)
Silver	Argentite	1,282,080	225.0	1.9
Sodium	Salt	11,492,856,000	4,743.3	39.9
Strontium	Celestite	13,142,810	1.3	<0.1
Tantalum	Tantalite	34,437	395.9	3.3
Tellurium	Tetradymite	9,129	485.9	4.1
Tin	Cassiterite	21,861,500	221.9	1.9
Titanium	Ilmenite and rutile	282,730,389	47.1	0.4
Tungsten	Scheelite	3,529,340	624.3	5.2
Vanadium	V in several ores	2,170,733	54.5	0.5
Yttrium	Monazite	111,93	0.4	<0.1
Zinc	Sphalerite	533,056,000	314.7	2.6
Zirconium	Zircon	47,442,300	739.2	6.2

potassium and copper. At the end of the list, we find elements such as germanium, strontium and yttrium, essential in certain specific applications.

The loss of total mineral wealth of the last 118 years, that is, that "bonus" lost that nature has given us by concentrating the elements in the crust reaches, at least, about 200 Gtoe for the minerals included in the table, which would be the theoretical amount of energy needed to replace or recover those mineral deposits from a dispersed environment with current concentrations and technology. Considering that the world oil reserves known today are 244 Gtoe, we would need to use almost 82% of those reserves to recover these minerals from Thanatia. It is important to note here that the exergy replacement costs have been theoretically obtained and that, as seen in Sect. 4.3, the true values will be significantly higher.

With this in mind, the annual theoretical degradation of mineral wealth is 38 Mtoe/year on average, with large variations between minerals. Aluminium, for instance, is at 940 Mtoe/year, whereas molybdenum carries a loss of "only" 1.5 Mtoe/year. These figures have been calculated taking into account the extraction from 1900 to 2018, but if we only consider the last few decades, the figure increases considerably due to the almost exponential trend of mineral extraction.

Knowing the accumulated extraction in this period of time (which has already been extracted) as well as the estimated resources of each mineral (which remains in the Earth's crust)—already seen in Sect. 3.3—the depletion rate can also be evaluated.

This method has its limitations: on the one hand, it must be noted that the resource figures available today are estimates deriving from data by mining companies, articles and scientific reports, as well as information from studies carried out by national geological services. On the other hand, a portion of these resources will never be extracted due to technical, environmental or social issues. Furthermore, it is assumed

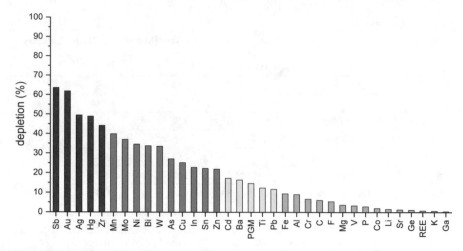

Fig. 5.3 Percentage of exhaustion of different elements based on the resources known today and the cumulative extraction from 1900 to 2018

that extraction will remain constant in the future and that there will be no new discoveries leading to variations in the resource figures. There are also some minerals whose resources are extremely large, such as limestone, sodium, silica and gypsum. Besides, and as commented previously, in some cases, there is insufficient extraction or resource data to reliably calculate the percentage of depletion: this applies, for example, to tantalum, tellurium and niobium.

Bearing this in mind, the results of this analysis can be seen in Fig. 5.3. The colour scale indicates the elements whose depletion rate is greater than 40% (in red) and those whose depletion rate is less than 10% (in green).

The elements that show a higher rate of depletion are antimony, gold and silver; elements located at the opposite end include potassium, strontium or magnesium. Aluminium, which as seen previously has the greatest loss of associated mineral wealth of all the elements, is located in the lower-middle part of the scale. This is because, despite being extracted in large quantities, the amount of available resources is very high, so it would still be available for many years to come providing the current extraction rate remains constant.

There are many other elements here that are also striking, including mercury, an element with one of the highest exhaustion percentages. However, given the restrictions on its use and the overall decrease in its extraction, the situation cannot be described as worrying. The case of lithium, located at the other end of the scale, is also notable among the elements that present the least amount of depletion. That said, it must be remembered that lithium extraction has increased eightfold in just over 10 years and that its future demand is certain to grow exponentially. Its status could therefore undergo very rapid changes. A similar situation can be observed in the case of rare earths, germanium or gallium. To be noted is that the available resource estimates of these last two elements, germanium and gallium, are very imprecise since they are extracted as by-products from other minerals.

Furthermore, data on the depletion rate can be combined with current extraction figures to calculate how long minerals will remain available provided the extraction rate remains constant. This rate is known as the R/P ratio in the literature, where R represents reserves and P is the extraction figure of each mineral. This same exercise can be carried out with resources, which as we have seen tend to be higher, although they include material that may never be possible to extract. Even if this approach has been applied fundamentally to fossil fuels (World Energy Council, 2016), it can also be applied to minerals. However, as this is a value that usually remains unchanged over time, it is foreseeable that variations in consumption trends—significant in the case of elements required in emerging technologies—new discoveries or estimates of available reserves, economic changes, etc., generate drastic changes in the figures in short periods. Thus, although these figures can be calculated theoretically, insights derived from them should be taken as a warning rather than as true indicators of criticality.

In the case of the R/P ratio calculated with the reserve data that we have seen in Table 3.1, the elements whose currently known reserves would be depleted in the next 20 years if the extraction figures from 2018 remain constant would be chromium, tin, gold, lead and zinc (Table 5.2). However, we have seen that these reserve figures

Table 5.2 Hubbert peaks for the various elements (Calvo et al., 2017a)

Element	Peak	R^2	Element	Peak	R^2
Aluminium	2084	0.9806	Magnesium	2192	0.7899
Antimony	2012	0.7499	Manganese	2030	0.8408
Arsenic	2059	0.3345	Molybdenum	2030	0.9372
Baryte	2080	0.9020	Nickel (late.)	2032	0.9409
Beryllium	2247	0.3887	Nickel (sulph.)	2033	0.9413
Bismuth	2040	0.7842	Gold	2014	0.8213
Cadmium	2082	0.9011	Palladium	2073	0.9518
Cobalt	2142	0.8983	Silver	2022	0.7066
Copper	2072	0.9877	Platinum	2075	0.9480
Chromium	2107	0.9576	Lead	2128	0.7962
Fluorspar	2153	0.8774	Phosphate rock	2181	0.9114
Tin	2086	0.7063	Tantalum	2039	0.8319
Gallium	2068	0.9031	Tellurium	2062	0.4670
Germanium	2236	0.7245	Titanium (rutile)	2082	0.8653
Graphite	2148	0.9235	Titanium (ilmenite)	2084	0.9604
Iron	2091	0.8807	Vanadium	2124	0.8909
Indium	2032	0.9851	Zinc	2061	0.9750
Lithium	2037	0.9190			

represent what is profitable to extract at present, and therefore cannot be taken as reference values. If we were to use resource data as a starting point instead of reserve data, the availability of lead would increase from 19 years to over 450 years. Resource data consist of much larger figures, so the value obtained may be considered more optimistic, but not always more realistic, as we will see below using chromium as an example.

Although chromium is present in various minerals, it is mainly extracted from chromite. It is produced in more than 20 countries, yet 80% of its extraction is concentrated in only 4: South Africa, India, Kazakhstan and Turkey (USGS, 2019). In the case of reserves, the estimates are 560 million tons; in terms of resources, estimates are around 12 trillion tons, located mainly in South Africa and Kazakhstan. Assuming the estimates of the resource-to-extraction ratio are correct, there would be enough chromium to last until after 2350 if the current extraction rate remains constant over time. However, if we look at the annual extraction of chromium (Fig. 5.4) in the last 20 years, we can see that it has increased threefold, and expectations are that its demand will continue to increase significantly. It is therefore not surprising that the European Commission pointed to chromium as critical in its 2014 report (European Commission, 2014). Still, in the last two reports, 2017 and 2020, chromium is no longer part of the EC critical raw material list (European Commission, 2017, 2020).

This R/P ratio, used to estimate the future availability of minerals, as can be seen, generates rather unoptimistic results, in some cases indicating that we will run out of certain elements very soon. As these are static data, they do not reflect current reality. We will therefore look into another method that *does* take into account the evolution of extraction over time in the next section.

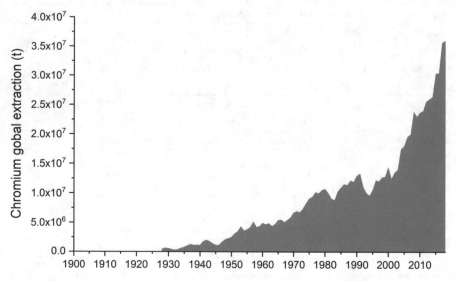

Fig. 5.4 Global extraction of chromium from 1900 to 2018. Figure developed with data from the USGS

5.2 Exergy Evolution of the Future Extraction of Minerals

Once the exergy evolution of the historical extraction of minerals has been analysed, the future situation can be contemplated using projections based on the Hubbert model, developed in the 1950s and based on the analysis of US oil extraction, also known as the peak oil theory.

The peak oil theory arose when Marion King Hubbert, a geophysicist who worked for Shell, analysed the extraction of oil fields from different parts of the USA (Hubbert, 1956). When studying the extraction patterns, he concluded that it was undergoing an evolution that resembled a Gaussian bell, beginning to rise almost exponentially until it reached a maximum point, and when extraction reached that point, it began to lower again, generating bell-shaped extraction curves (Hubbert, 1962).

This model developed by Hubbert is based not only on extraction data but also on data on reserves and/or available resources, data will determine the size of the area below the curve. First, it is to be noted that no extraction with an exponential tendency can remain in its current state for a long time, since the physical limits, reserves and resources will cause this extraction to decrease after a certain amount of time. For any extraction curve with a finite amount of resources, two points are known from the beginning: $t = 0$ and $t = \infty$, where the extraction rate will be zero in both cases since the resource has ended at $t = \infty$. It is also known that this curve will have between one or more maximum extraction peaks.

Second, it must be considered that the area under the curve must be the sum of all available resources. Therefore, the maximum theoretical peak of extraction is reached when all the highest quality and most accessible resources have been extracted. This extraction decreases as less quality and less accessible resources are extracted until the moment the resource is exhausted (at $t = \infty$). The shape of the extraction curve will approximate a symmetrical bell (Fig. 5.5) that can be adjusted using the following equation (Eq. 5.1):

$$f(t) = \frac{R}{b_o \sqrt{2\pi}} e^{-\frac{1}{2}\left(\frac{t - t_o}{b_0}\right)^2}$$

(5.1)

Fig. 5.5 Bell shape of the Hubbert model for the extraction cycle of any natural resource

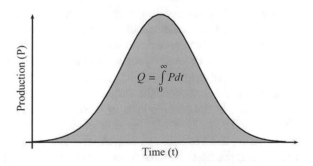

$$Q = \int_0^\infty P\,dt$$

Production (P)

Time (t)

where R is the quantity of reserves or resources of the element to be considered, either fossil fuels or minerals, and where the unknowns are the parameters b_0 and t_0.

The maximum of the function will be determined by t_0, which meets the following (Eq. 5.2):

$$f(t_0) = \frac{R}{b_o\sqrt{2\pi}} \tag{5.2}$$

For any prediction made using this model to be reliable, the resource figures, which are a determining factor, must be as accurate as possible (Höök et al., 2010). Furthermore, even if these initial data are reliable, they do not follow that these resources may be exploited at some point in their entirety, either for socio-political or technological reasons. In fact, actual extraction curves tend to be asymmetric, with a decline that is much steeper than the increase as different economic, geological, political and technological factors come into play. It is, therefore, necessary, prior to its application, to clarify the limitations of this model, particularly if one intends to use it in the case of minerals.

In total, six types of limitations can be considered, related to: (1) the inadequate precision of resource data, (2) environmental issues, (3) socio-political issues, (4) whether we are dealing with primary products or by-products, (5) substitution and recycling and (6) changes in the concentration of minerals.

Since the entire upper crust of the Earth has not been thoroughly explored, the resource figures estimate the available amount of a specific material; estimates can be calculated both at national and global levels; and are usually constant over time. Besides, these estimates do not indicate exactly the maximum amount of the element that could be produced—they are estimates based on current geological knowledge, since there will always be a portion that is inaccessible, unprofitable or technically impossible to extract. This information usually derives from reports of mining companies and data that different organisations collect until global figures are calculated. Yet, these figures are based on the economy, since the sale price of the commodities will influence the number and type of exploration campaigns carried out by mining companies (Tilton & Lagos, 2007). Furthermore, this exploration is not usually carried out in a uniform way. For instance, some metals attract more attention than others. Examples include valuable metals such as gold or platinum, or elements whose supply can undergo drastic changes due to political factors, such as rare earths. Consequently, this is reflected in the rates of discovery of new deposits and in the efforts that different companies focus on each specific element.

Furthermore, some natural resources have a negative impact on the environment or on health, such as arsenic, mercury or cadmium. For this reason, due to the sudden changes in their extraction, the Hubbert model is not useful in producing representative results. In the specific case of mercury, extraction gradually increased from 1900 to the 1970s. However, because of the drop in prices and the restrictions on its use owing to its toxicity, its extraction reached historical lows in 2010. In recent years, the extraction of mercury has experienced a rebound, with China being the main producing country, responsible for almost 90% of global extraction in 2018.

This is due to China being a country where environmental restrictions are laxer than in other countries. Even so, the figures are still very different concerning previous extraction data (USGS, 2019). Leaving aside the fact that, in the future, interest in this element will most likely see a decrease; its exhaustion date cannot be reliably estimated.

Another important fact is that, as mentioned, Hubbert's theoretical curves are symmetrical, but some socio-political factors can influence extraction, including conflict, political instability in one of the main producer countries of a commodity, shortages in international supply, etc. For example, an element whose extraction relies heavily on market price is gold. There is also the case of rare earths in China (explained previously), whose export restrictions in 2011 caused prices to rise sharply, in turn stimulating mining exploration in other regions of the planet (Wübbeke, 2013). These alterations can generate different peaks in extraction, meaning that the ideal Gaussian bell shape of the Hubbert model does not always coincide with the true situation on the ground.

It should also be noted that, although there are some elements that can be obtained directly in many mines, the vast majority of them are obtained as a by-product of others. Thus, it is not so common for the extraction of these by-products to follow the bell pattern since extraction will inevitably be linked to that of another element. An example is indium, which is used in photovoltaic panels and liquid crystals, and which is usually extracted as a by-product of the processing of zinc minerals and copper; in addition, since the resource data for said by-products are calculated based on the resources of the main elements together with those in question, data reliability inevitably decreases.

Finally, the Hubbert model neither takes into account recycling nor substitution, factors that can also significantly alter extraction curves. However, it is also necessary to point out that many elements that are used in technology today are not easily recoverable or recyclable, but this could change in years to come.

Certain studies have shown that the average ore grade for some metals is beginning to decrease worldwide and, as this grade decreases, so does the quality of the ore, exponentially increasing the energy costs associated with mining and processing (Mudd et al., 2013; Mudd, 2014; Mudd & Jowitt, 2014). As a direct consequence, larger amounts of water are needed, more tons are processed and more mining waste is generated. Because of this variation in ore grade, when the Hubbert model is applied to minerals, the exergy (or rather, the exergy replacement costs) instead of a ton-based model, better reflects the situation by taking into account resource quality as well as quantity. While the quality of fossil fuels remains constant with extraction, this is not the case with minerals. The maximum theoretical peak of extraction in the case of minerals is reached when the highest quality resources have been extracted, and the extraction begins to decrease as those of poorer quality or with lower concentrations are extracted. Furthermore, when using exergy instead of tons, it is easier to represent the curves of different elements in the same diagram, since the orders of magnitude are similar.

Taking these caveats into account, using the Hubbert model combined with the exergy replacement costs, we have calculated the maximum extraction peaks for the different elements. The accumulated extraction from 1900 to 2015 and the estimated resources (Calvo et al., 2017a) have been considered. Some data come from the USGS, as we have seen in Table 5.2, but in the case of certain specific elements, we have had to resort to other bibliographic sources, which included estimates based on different criteria (geological, mining, etc.).

In contrast to the previous two methods based purely on static values (LMW and R/P), this current method allows us to consider the evolution of production, as its trend is incorporated in model calculations. It should be noted that, to calculate the curves, the reduction of the concentration in the mines over time and extraction has not been considered. The extraction curve is, therefore, equivalent to that of the exergy replacement costs. In order to carry out more precise estimates, we would also need to consider this concentration reduction. The advantage of using "static" replacement exergy costs instead of extraction data is that, as explained previously, the curves can be represented in the same diagram.

If the extraction rate of all the elements is analysed, there are six that stand out from the rest due to the large amounts extracted annually. These elements are aluminium, chromium, copper, iron, manganese and zinc. The maximum peaks of extraction of these six elements are indicated in Fig. 5.6.

The metal with the closest peak of maximum extraction is manganese, which in theory would be in 2030, and the furthest is that of chromium, which would take place in 2107. As we saw in the previous section, using resource data to estimate the R/P ratio of chromium, and assuming that the 2018 extraction rate remained constant in the future, chromium would remain available for many more years, up

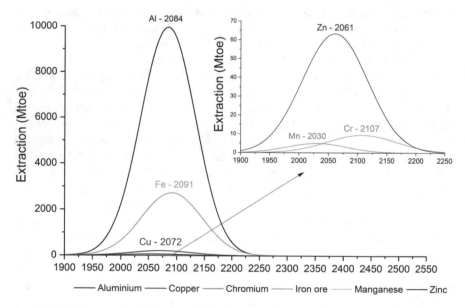

Fig. 5.6 Bell shape of the Hubbert model for the extraction cycle of any natural resource (Valero & Valero, 2014)

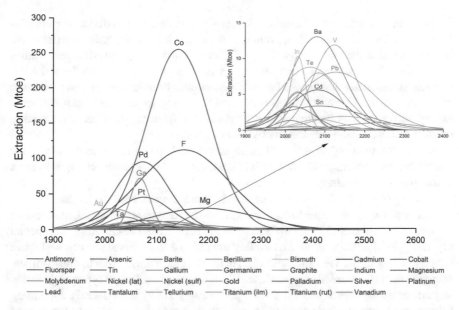

Fig. 5.7 Hubbert peaks for different elements considering the total resources available (Calvo et al., 2017a)

to and beyond 2350. In this case, taking into account the evolution of production, which we have already seen accelerating, we could experience issues with chromium supply much earlier. This does not imply that the entire supply of chromium will be exhausted by 2107; it simply means that from that point the amount of chromium that is extracted will be less. The quality of the mines in operation at that time will also have decreased.

So far we have seen the maximum extraction peaks of the six main elements, but there are many other elements that can be analysed, the results of which are indicated in Fig. 5.7.

According to these data, if the extraction continues with the trends of the last decades, the future availability of some elements will be at stake. In the specific case of gold, its maximum extraction peak would have occurred in 2014 according to currently available information, which coincides with other studies that also observe that the peak has already occurred or will occur in the coming years (Sverdrup & Ragnarsdottir, 2014). Antimony is another element whose maximum extraction peak would have already passed. In the case of both elements, however, the information related to resources is very limited and sometimes incomplete, so the data should be taken with caution. A clear decrease in the extraction itself is not yet observed as it has not been long since this supposed maximum peak occurred.

Furthermore, there are a total of 12 elements whose maximum extraction peak is expected in the next 50 years, including bismuth, indium, lithium and nickel, and if we consider the next 100 years, this figure increases to 30 elements (Table 5.2).

In all mathematical calculations, it is advisable to understand the degree of reliability of the data, which is why the values of the coefficient of determination (R^2) have been included in the table. Thus, the quality of the maximum extraction peak values obtained can be evaluated so that low values of R^2 represent data with limited reliability, as is the case of arsenic and beryllium. A similar situation can be observed in the case of tellurium. This element has been used for a very few decades and is extracted as a by-product of the treatment of electrolytic slimes from copper refineries, meaning reliable resource data is almost impossible to access.

In addition to applying this method to all minerals using resource data from the United States Geological Survey (USGS), we can also analyse the changes in peak extraction if changes in resource data occur.

Lithium is a mineral considered critical in many studies because its demand will experience great growth due to its use in electric car batteries. Today the main use of lithium is in batteries (35%), but lithium oxide is also used in ceramics and special crystals (USGS, 2019). As we are trending towards a decarbonised society, we expect the demand for lithium to increase tenfold in the next 30 years in order to transform the transport fleet (García-Olivares et al., 2012). It appears in different types of deposits, in particular pegmatites and brines. In pegmatite formations, minerals containing lithium are mined, with Australia being the main producing country. In the case of brines, Chile is the main producer, possessing over half of the world's known lithium resources. The Salar de Atacama, with the highest known concentrations of lithium and potassium, is one of the most relevant deposits globally.

A number of studies seek to estimate the amount of lithium resources available on the planet, one of the most common being carried out by the United States Geological Survey (USGS, 2019). Since 2000, the estimates of lithium resources have tripled, from 12.76 million tons to 40 million tons (USGS, 2016). Other estimates put these resources between 31 and 71 million tons, the highest figure in the bibliography being 116 million tons (Gruber et al., 2012; Kesler et al., 2012; Mohr et al., 2012; Sverdrup, 2016). In all these cases, the amount of lithium that could potentially be extracted from seawater is not considered. We can obtain different values of maximum extraction peak (Fig. 5.8) by applying the Hubbert model. In the most pessimistic estimates, with only 31 million tons (Kesler et al., 2012), this peak would take place in 2032; in the most optimistic estimates—116 million tons (Sverdrup, 2016)—the maximum peak of extraction would not occur until 2060. Even if we were to multiply this last figure by two, the peak would only gain 18 more years in the future.

Another possibility would be to try to extract the lithium contained in the seawater, whose average concentration is 0.17 ppm. This would give us a theoretical value of 230 billion tons of lithium present in seawater, of which approximately 20% could be recovered (Yaksic & Tilton, 2009). Compared with brines and lithium-containing minerals, seawater appears to be a much more promising source and could go a long way in meeting expected demand. Considering this potentially recoverable amount of lithium, the peak of maximum extraction would take place in 2246, hundreds of years after the previous peaks.

There are some examples of countries that have tried to recover lithium from water in semi-industrial plants. Conventionally, lithium can be extracted from water

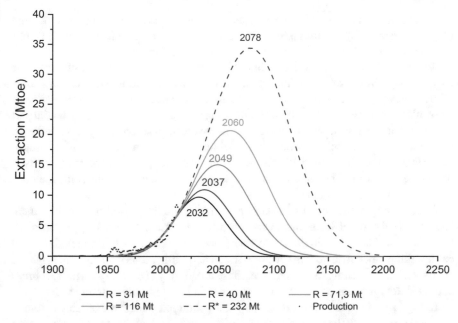

Fig. 5.8 Different values for the Hubbert peak of lithium based on different resource data (Calvo et al., 2017a)

through a co-precipitation and extraction process with an ion exchange and sorption process, but there are also other techniques such as the extraction of liquids or the use of different types of membranes (Swain, 2017). A Japanese experiment carried out in the 1980s managed to recover 750 g of lithium carbonate after treating almost 4200 m^3 of seawater, with an efficiency of 19.7% (Fasel & Tran, 2005). In South Korea, the firm POSCO is carrying out a project to extract lithium from seawater, using a system that is 30% more efficient than previous methods. However, today this technology is not feasible on a large scale, since it requires the processing of large amounts of water to obtain significant amounts of lithium, with the associated energy consumption.

In addition to lithium, phosphorus is another of the most relevant elements for the green economy, whose availability can put at risk sectors as important to the population as agriculture. Phosphorus is one of the six key elements that make up the Earth and life on it. Phosphate rock is the main source of phosphorus on the planet and is made up of hundreds of phosphorous-containing minerals, most notably apatite (calcium phosphate) that can be exploited commercially with or without processing. There are two main types of phosphate rock deposits, sedimentary and igneous, with the first type being the more commonly exploited. The mines in which it is extracted are usually both open pit and inland, with an extraction process very similar to that of coal.

Almost 95% of the world's phosphorus extraction is destined for the manufacture of fertilizers and food, with the remaining 5% used in some industrial applications that include water treatment, electronics, lubricants, medicine, etc.

Today, China is responsible for almost half of the global extraction of phosphate rock followed by Morocco and the USA, with Morocco boasting over 70% of known world reserves (USGS, 2016). The vast majority of the phosphorus that is used comes from this primary extraction, since it is not being recycled. However, a substantial part of manure or other agricultural residues could be recovered. There are few studies on available phosphate rock resources, but the figures range from 70,000 million tons to over 300,000 tons (USGS, 2019; Van Kauwenbergh, 2010). This great variability is due to the fact that neither all companies calculate resources using the same methods nor all data similarly reliable.

In the specific case of USGS data, a source frequently used worldwide, there have been substantial variations in resource figures in recent years, going from 16 Gt in 2010 to 300 Gt in 2019. These variations are due, in part, to the updates and exploration work carried out in several African and Latin American countries, but we encounter here the same issue as before: these are merely estimates of what resources could exist, regardless of whether or not they can be profitably extracted.

In Fig. 5.9, the maximum extraction peaks obtained with the available resource data and also with 2019 reserve data of 70 Gt have been represented for comparison purposes. It can be seen that in the case of the 2010 resource figures (16 Gt), the maximum theoretical peak of extraction would take place in 2033, which would

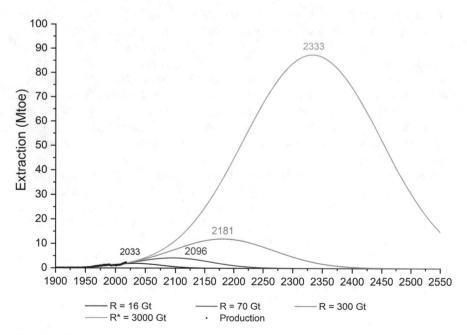

Fig. 5.9 Different values for the Hubbert peak of phosphoric rock based on different resource data

move further away in time by a few years if this amount was higher. As explained above, reserve data are much more dynamic than resource data, nonetheless it helps us compare with the more optimistic resource data, in this case from USGS (300 Gt), producing the maximum peak of extraction in 2181. This point in time is far into the future—it is unlikely that future generations will experience geological supply issues of phosphoric rock, should demand match expectations. The supply issues that might arise would then be more related to socio-political aspects and that is precisely the reason why phosphate rock is considered critical by some institutions such as the European Commission.

Given that the value of 300,000 million tons of phosphorous resources from the USGS might be an underestimation, we have also calculated the maximum extraction peak were this figure to be multiplied by ten. Under this hypothesis, the maximum peak of extraction would take place approximately 150 years later. In other words, we would be talking about having access to sufficient phosphorous until the first half of the twenty-fourth century, a time scale that many of us cannot even imagine.

Another option, both in the case of lithium and phosphorus, is to attempt to extract them from the crust (of Thanatia) rather than from the deposits and places where they are already concentrated, resorting to a much more dispersed source. Depending on the average composition of the crust, there are between 20 and 24 ppm of lithium and 700 and 1200 ppm of phosphorus (Jaupart et al., 2004). Current quantities of lithium and phosphorous are certainly much greater than any number of resources available today, but the technology and the quantity of material to be processed and the energy required for this would be so high that the process would be almost impossible to carry out.

Although Hubbert's model may have some defects, as seen previously, it can serve to give a rough idea of how long we can maintain the current extraction rate and at what point this will start to decrease. Logically, these figures will vary depending on previously mentioned factors, but they can still help to inform policymakers on materials management, to promote recycling or the substitution of more critical materials with other less critical ones.

5.3   The Mineral Balance of Countries and Regions

Analysing the loss of mineral wealth on the planet, that is, the mineral depletion caused by primary extraction, is a task that can be carried out not only globally as we have seen in the first section of this chapter but can also be applied to a specific country or region to determine the quantity and quality of the extracted mineral, in terms of mass and exergy. This study can also be complemented with an economic analysis, comparing it with the GDP of the mining sector, that is, contrasting it with the profits generated by the sale of the same mineral extracted.

To carry out this analysis, Sankey-type flow charts are very useful for determining the inflows into the system, that is, the sum of imports and primary extraction, and outflows: exports, consumption and the part corresponding to recycling. In these diagrams, the width of the arrows of each input and output is proportional to flow quantities, allowing us to determine material transfers easily and visually within the country or region being studied.

Among the inflows is primary extraction, information that can be obtained from the mining statistics of each country, from geological services (USGS, BGS) or from a variety of national services. Data on imports (inputs) and exports (outputs) can be more difficult to obtain in the case of minerals, since in many cases, the aggregated information appears in the form of metallic and non-metallic minerals, but there are national agencies and chambers of commerce that do provide this information categorised by minerals. Recycling, which could conceivably be considered both an input and an output, is considered as an output in this study since we can assume that the amount of material produced in a specific year cannot be recycled that same year. The recycling figure data for each metal have been taken from UNEP (UNEP, 2011). Finally, domestic consumption has been calculated as the subtraction of inputs and outputs.

Now that the basis of this section has been explained, three practical cases will be analysed: (1) Spain and Colombia, (2) Latin America and (3) Europe. These regions have been chosen for their representativeness, since, even if historically it was a country with a large mining industry Spain is fundamentally an importing region. This same situation can be observed in Europe. The opposite occurs in Latin America, a region well known as a supplier of raw materials. It is, therefore, appropriate to include a case at both ends of the scale. Since statistics vary by country and region, the most recent year for which sufficient information was available to carry out the analysis has been used. It should be taken into account that national statistics are not always released year after year, and that in many cases, there is a time delay. Therefore, in the case of Colombia, the year for which complete information was available is 2011—for comparison purposes, the same year has also been chosen for Spain. Data from 2013 have been taken as reference for Latin America, and data from 2014 have been taken as reference for Europe.

5.3.1   Spain and Colombia

Spain is a country whose mining industry is fundamentally based on the extraction of industrial minerals: in 2011, for example, Spain was the third-largest producer of gypsum worldwide and the sixth-largest producer of fluorspar. Still, there are some active metallic mineral mines today. The total contribution of the mining sector to GDP was 0.8%, a rather low value when compared with other countries (Gurmendi, 2013). Despite the fact that Spain does not have proven oil or natural gas reserves, it does have 530 million tons of coal, of which around 6.7 million tons were extracted in 2011 (British Petroleum, 2014).

Figure 5.10 shows the mineral balance of Spain for 2011, represented in tons (on the left) and in exergy replacement costs (on the right). The legend only includes

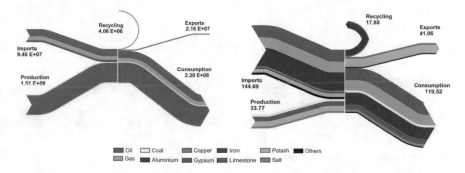

Fig. 5.10 Sankey diagram with inputs and outputs for Spain for 2011. Data in tons (left) and exergy replacement costs (right) (Calvo et al., 2015)

those minerals that can be seen with the naked eye in the figure, but a total of 30 have been taken into account, in addition to fossil fuels.

In tons (diagram on the left), limestone clearly prevails over the rest of the materials, both in inputs (import and extraction) and outputs (consumption, export and recycling). In this case, limestone represents more than 86% by mass of the total extraction of minerals in Spain in 2011, with almost 130 million tons extracted. This figure is certainly very high when compared with the 8 million tons of gypsum extracted the same year, the second most extracted mineral by weight. Imports, with a lesser relevance than extraction, are mainly oil, natural gas and bauxite, although, in the case of the latter mineral, the amounts are negligible.

If this mineral balance is expressed using exergy replacement costs (on the right in the diagram), we can observe that the weight of the limestone decreases drastically in the case of extraction, down to 23% of the total national extraction. The explanation is that limestone is a very easy mineral to extract and very abundant in the Earth's crust, so its replacement cost per ton is very low. Likewise, the change that takes place in exports, where potassium would predominate, or in recycling, where aluminium becomes visible, can also be seen. Although at first glance, it may appear that Spain is a self-sufficient country in terms of mineral resources, the data show that, in reality, it is highly dependent on imports.

If we look at Colombia, in 2011, approximately 40% of the territory had been licensed or requested for mining concessions (Sanchez-Garzoli, 2012). In addition, Colombia was Latin America's fifth economy that year, the contribution of the mining sector being 11.3% of GDP, 55% of income coming from exports.

In 2011, Colombia was the first coal-producing country in Latin America and the fifth globally. With regard to minerals, Colombia is a country with a great mining tradition: it is the world's leading producer of emeralds, the tenth-largest producer of gold, the largest producer of nickel in South America and the sole producer of platinum in Latin America (SIMCO, 2011). If we look at the country's mineral balance for 2011 in tons (Fig. 5.11), primary extraction and export clearly predominate over the rest of the flows in the system, representing almost 92% of the national extraction of fossil fuels compared with other minerals. Eliminating the data on fossil fuels and

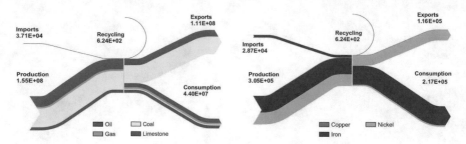

Fig. 5.11 Sankey diagram with inputs and outputs for Colombia for 2011 (data in tons) (Calvo et al., 2015)

limestone in the figure (diagram on the right) allows us to highlight those metals that Colombia produces in significant quantities, such as nickel and iron. For example, we can see that practically all the nickel that is extracted in Colombian territory is destined for export.

Let us now look at this same mineral balance expressed in exergy replacement costs (Fig. 5.12). Again, fossil fuels continue to have the greatest weight, but if they are removed from the balance (diagram on the right), we can see the fundamental role played by other elements that could not be appreciated in 2011. Gold, which in Fig. 5.11 could not even be represented due to its low weight in mass, is clearly highlighted against other metals such as nickel, which could be appreciated in the figure expressed in tons. This is due to the fact that gold is an element whose concentration in the Thanatia crust and in the mines differs greatly, and whose extraction and processing process are very energy intensive.

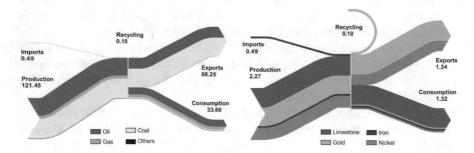

Fig. 5.12 Sankey diagram with inputs and outputs for Colombia for 2011 (data on exergy replacement costs) (Calvo et al., 2015)

5.3.2 Latin America

After studying a country that bases its consumption on metals' importation, such as Spain, against another country where the opposite occurs, Colombia, the analysis can be extended to a region that is the main supplier of metals and natural resources to other countries: Latin America.

Specifically, primary extraction, imports and exports of 20 countries in Latin America and the Caribbean were analysed: Argentina, Bolivia, Brazil, Chile, Colombia, Costa Rica, Cuba, Dominican Republic, Ecuador, El Salvador, Guatemala, Honduras, Mexico, Nicaragua, Panama, Paraguay, Peru, Uruguay, Venezuela and Haiti. In each case, the statistical services of each country, where applicable, have been used as well as databases from the USGS, BGS (British Geological Survey) and OLADE (Latin American Energy Organization).

All these countries have very different characteristics in terms of population, economic situation or geological situation. It is the last variable that precisely conditions the availability of each element in the territory based on existing mineral deposits.

Latin America is known for its high extraction of fossil fuels, which in the case of oil represented almost 8% of the world total and 5% in the case of natural gas in 2018 (BP, 2019). However, in this territory, there are also numerous large mineral deposits, including copper, iron, zinc and silver deposits, among others. Niobium deposits in Brazil are also of great relevance worldwide, being the main global producer and also the country with the largest amount of reserves of this element. In the case of lithium, its extraction is distributed between Argentina and Chile, the latter being the country with the largest reserves of this element worldwide (USGS, 2016, 2019).

Figure 5.13 represents the extraction and the amount of reserves of each element in Latin America. Niobium and lithium, both critical in the production of electric

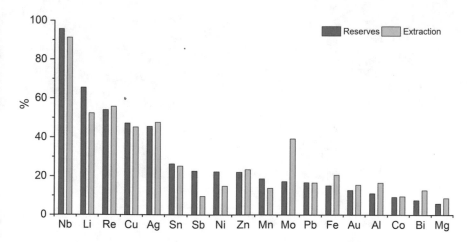

Fig. 5.13 Percentage of reserves and extraction in Latin America with respect to the global total (2016) (Palacios et al., 2018b)

cars and batteries, have already been discussed, but there are many other elements for which this region is an indispensable source, such as rhenium and copper (Palacios et al., 2018a).

In terms of silver, Latin America was responsible for almost half of the global extraction of this element, with Mexico and Peru being two of the most important producers. If we add to this that it takes just over 100 g of silver to manufacture a photovoltaic solar panel, the situation of green technologies could be jeopardised in the event of interruptions in supply.

Not only is it important to know the significance of extraction in Latin America for the different substances; exports are another key factor to be taken into account when analysing the loss of mineral wealth in the region. It is precisely that these sales make it possible to cover external demand that favours extractivism in the region.

Figure 5.14 shows the export flows from this region to other areas of the planet in 2013, specifically to North America, Europe, Africa, Asia and Oceania (Palacios et al., 2018b). A total of 38 minerals have been taken into account to produce this figure, although due to the scale, only those with non-negligible values appear in the legend.

Analysing the mass data, iron and aluminium were the most exported elements, much of it to Asia, followed by salt and copper. However, studying these same results, and applying exergy replacement costs, the importance of iron remains unchanged, but other elements such as gold, silver and nickel, become more relevant.

It is to be noted that approximately a quarter of all the minerals mined in Latin America ended up in North America in 2013, mainly in the USA, while the rest

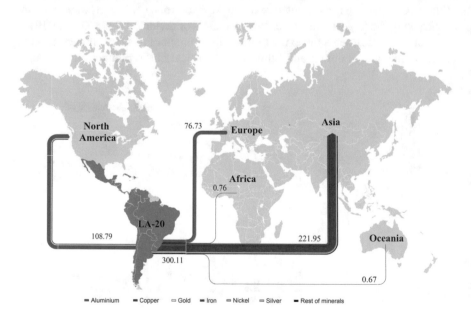

Fig. 5.14 Exports from Latin America (LA-20 in the figure) to the rest of the world regions for 2013 in exergy replacement costs (Palacios et al., 2018b)

was distributed among the other regions. This makes sense from the point of view of the proximity of this country with respect to the region studied, since proximity is a key factor in reducing the cost of transporting minerals. To this, we must add that the USA, in its mineral defence accumulation program for national defence, identified aluminium oxide and antimony, supplied by Venezuela and Mexico, as strategic minerals (U.S. Department of Defense, 2015).

Almost 38% of the copper produced in the region went to China, as did 48% of silver, while zinc went mainly to Russia (41%) and gold to Europe (21%). It is not surprising that China was one of the main receiving regions for raw materials, given its role as a supplier of products and goods to other countries. Furthermore, in EU criticality reports, we can see that Latin America, specifically Brazil and Mexico, was the main supplier of niobium and fluorite (European Commission, 2017).

This high export value also makes practical, as well as economic and political sense, since many companies that own mines in Latin America have processing and refining plants in other parts of the world to which they transport the material directly.

5.3.3 Europe

In the same way that the mineral balance of Colombia, Spain and Latin America has been studied, the same analysis can be carried out for the list of materials critical to the European economy from the point of view of exergy, supply risk and economic importance. On this occasion, instead of using replacement costs, the thermodynamic rarity has been used to show the practical application that both concepts have. Thus, by means of thermodynamic rarity, the true incorporated energy of each element is taken into account. It makes no sense to analyse fossil fuels separately, as was done previously using replacement costs, since double accounting would occur.

Comparing this mineral balance of Europe in terms of mass and thermodynamic rarity, we can observe several differences (Fig. 5.15). Although the 20 minerals listed in the legend are included in the diagrams, some of them, such as gallium, germanium or indium, are produced or consumed in such small quantities that the scale does not allow them to be visualised.

In mass terms, the main minerals imported by the European Union during 2014 were chromium, fluorspar and phosphate rock, which makes sense if we consider that the latter product is the basis of the agricultural sector and that European extraction is almost non-existent when compared with demand (Calvo et al., 2017b). The most extracted mineral was magnesite, monopolising almost 61% of the total internal extraction, which again is a widely used element in the agricultural and industrial sectors. As can be seen, most of the extraction is concentrated in industrial minerals, which are generally not usually exported or recycled, and which, due to their moderate commodity prices, are also consumed internally.

If this information is analysed using the thermodynamic rarity as a unit of measurement, expressed in Mtoe, the situation changes drastically, as we can see in the diagram on the right. In the case of imports, we went from seeing industrial minerals

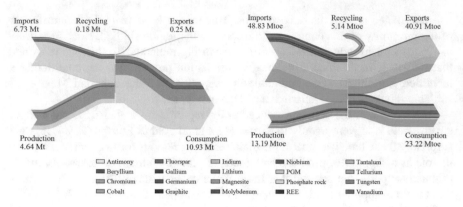

Fig. 5.15 Sankey diagram with inputs and outputs of critical minerals for the European Union for 2014 in mass (Mt, left) and in thermodynamic rarity (Mtoe, right) (Calvo et al., 2017b)

to metallic minerals such as the elements of the platinum group, tellurium, tantalum or niobium. In mass, this group of substances barely represented 2%, but when expressed in Mtoe, it represents almost 87% of total imports. Taking a closer look at this, almost all of these elements have also been considered critical by the European Commission, precisely because of the area's dependence on external sources. The large percentage difference that we find, in addition, helps us understand the criticality of these substances: using thermodynamic rarity, we not only take into account the scarcity in the Earth's crust and in mines (avoided costs) but also the energy necessary to extract these resources (energy costs actually incurred).

In the case of European extraction, once again the elements that stand out the most are cobalt, magnesite, fluorspar and tellurium, representing 95% of the extraction in Mtep (Calvo et al., 2017b). Magnesite, which in mass accounted for almost two thirds of total extraction, now barely reaches half. This decrease is related to the fact that it is a more common element than others in the crust, and that the energy required to extract and process it is not as high when compared with other elements. On the contrary, an element that in mass barely accounted for 0.5% of the extraction and that was not even seen in the figure—cobalt—in thermodynamic rarity, it increases to account for almost a third of total extraction.

Another significant change is the relative importance of exports. In mass, the arrow was barely visible and we could almost deduce that they were not very significant; however, when applying thermodynamic rarity, this increases considerably. If we add outputs from the system, exports, consumption and recycling, exports represent almost half. This great difference is due to the fact that the European Union exports tantalum, elements of the platinum group, tellurium, gallium and niobium, minerals whose thermodynamic rarity values (in GJ/t) are very high. Although in Europe, there are no mines that extract these elements, there are different metal processing and refining plants that obtain them as a by-product. For example, the Antwerp refinery in Belgium produces seven precious metals as a by-product, including platinum and palladium (Hagelüken & Meskers, 2010). Once these by-products are recovered, they are usually exported to other countries, even generating a deficit when the import is compared to export.

Analysing the mineral balance of Fig. 5.15 as a whole in mass terms, it can be noted that Europe is overall balanced in terms of minerals, since domestic extraction practically equals imports. However, if we consider element quality as well as quantity, we can see that Europe is actually importing much scarcer and more difficult to replace materials than the ones it is producing. In short, the European mineral balance is clearly in deficit.

5.4 Selling Cathedrals at Brick Price

The exergy replacement costs and thermodynamic rarity already give us enough information about mineral depletion, however, they can be combined with a series of economic factors to link economics and thermodynamics, this being one of the objectives of the environmental economy. By converting exergy costs into monetary costs, it is easier to interpret the data and compare it with other economic indicators of the mining sector, such as the gross domestic product associated with this sector. Carrying out this conversion should not be considered the final objective, since the physical information that the exergy replacement costs give us is more than valuable in and of itself.

In the case of fossil fuels, the corresponding monetary cost can be easily calculated with the sale price of these in the year under consideration. These sales prices, both historical and current, can be obtained from different statistics, such as those of British Petroleum (BP), which publishes an annual report containing this type of information.

In the case of minerals, the conversion is not so direct and it is necessary to consider a range of values, establishing both a lower and a higher boundary. This is required since exergy replacement costs represent the amount of energy needed to restore mineral wealth with the current technology available, a process that can be carried out using different energy sources.

Let us now analyse what the monetary cost of extraction would have been in exergy terms considering two sources of energy to replace the mineral wealth extracted in each country. The cheapest source of energy to reconcentrate these minerals is coal, and the most expensive is electricity (Calvo et al., 2015).

In order to carry out this exercise, data have been taken from 2000 to 2012 for Spain and Colombia, given that they are very disparate but highly representative regions in terms of primary mineral extraction; Fig. 5.16 shows the results obtained.

In Colombia's case, the GDP generated by the mining sector in that period of time is clearly lower than the reconcentration of this extracted mineral wealth would be, equivalent in monetary costs. In other words, the sale price of minerals is not sufficient to compensate for the loss of Colombia's mineral wealth, and these should be considerably higher for full compensation to occur.

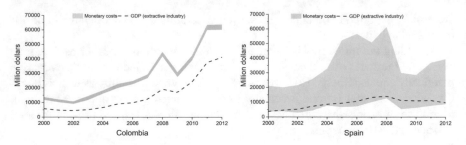

Fig. 5.16 Monetary costs of the mineral patrimony loss and GDP of the mining sector for Colombia (left) and Spain (right) (Calvo et al., 2015)

Even for Spain, the mining sector's GDP barely makes up for the loss of the country's mineral wealth. In fact, if the upper limit was considered, that is, if the energy source used to replace the extracted mineral resources was electricity (the most expensive energy source), the monetary costs associated with the replacement would be much higher than those benefits obtained from the sale of resources.

If this same analysis is carried out with the data previously obtained from Latin America, we see that on average, the recovery of these metals would be between one and three times greater than the economic benefit obtained from their sale, that is, the profits were not even close to offsetting the loss of mineral wealth in Latin America (Palacios et al., 2018b).

These results rely heavily on the prices corresponding to the different energy sources, which may rise or fall depending on factors outside the system itself. However, they allow us to assign an order of magnitude to the loss of mineral wealth in economic terms. The mines have taken thousands of years to form and the amount of ores present in them is not infinite. Let us suppose a tower made up of 7,300 tons of iron; taking into account the average sale price of iron in the market (about 81 USD/t), this could be sold at just over $590,000 (around 540,000€ at the time of publication). However, it is unlikely that anyone would think that this is a reasonable price to sell the Eiffel Tower. By extrapolating this example to mineral resources, Latin America is selling its natural "monuments" at absurdly low prices. The mineral cathedrals that were created over millions of years, which today constitute the natural heritage of many countries, are literally being sold at the price of bricks. What will future generations think about us mortgaging their future? This situation is clearly unsustainable and unfair, and the problem, far from decreasing, seems to be getting worse.

References

BP. (2019). Statistical Review of World Energy 2019. British Petroleum.
British Petroleum. (2014). *Statistical review of world energy*. BP global, London, UK.

Calvo, G., Valero, A., Carmona, L. G., & Whiting, K. (2015). Physical assessment of the mineral capital of a nation: The case of an importing and an exporting country. *Resources, 4*(4), 857–870.

Calvo, G., Valero, A., & Valero, A. (2017a). Assessing maximum production peak and resource availability of non-fuel mineral resources: Analyzing the influence of extractable global resources. *Resources, Conservation and Recycling, 125.*

Calvo, G., Valero, A., & Valero, A. (2017b). Thermodynamic approach to evaluate the criticality of raw materials and its application through a material flow analysis in Europe. *Journal of Industrial Ecology.* https://doi.org/10.1111/jiec.12624.

European Commission. (2014). *Report on critical raw materials for the EU. Report of the Ad hoc working group on defining critical raw materials.*

European Commission. (2017). *Study on the review of the list of critical raw materials. Critical raw materials factsheets.*

European Commission. (2020). *Study on the EU's list of critical raw materials (2020). Final report.*

Fasel, D., & Tran, M. Q. (2005). Availability of lithium in the context of future D-T fusion reactors. *Fusion Engineering and Design, 75–79,* 1163–1168.

García-Olivares, A., Ballabrera-Poy, J., García-Ladona, E., & Turiel, A. (2012). A global renewable mix with proven technologies and common materials. *Energy Policy,* 561–574.

Gruber, P., Medina, P., Keoleian, G., Keslser, S., Everson, M., & Wallington, T. (2012). Global lithium availability, a constraint for electric vehicles? *Journal of Industrial Ecology, 15*(5), 760–775.

Gurmendi, A.C. (2013). *The mineral industry of Spain. Years 2006 to 2012.* United States Geological Survey.

Hagelüken, C., & Meskers, C. E. M. (2010). Complex life cycles of precious and special metals. In V. der V. Graedel (Ed.), *Linkages of sustainability* (pp. 163–197). MIT Press.

Höök, M., Bardi, U., Feng, L., Pang, X., Bardi, U., Feng, L., & Pang, X. (2010). Development of oil formation theories and their importance for peak oil. *Marine and Petroleum Geology, 27*(9), 1995–2004.

Hubbert, M. K. (1956). *Nuclear energy and the fossil fuels.*

Hubbert, M. K. (1962). *Energy resources: A report to the Committee on Natural Resources of the National Academy of Sciences,* Washington, D.C. (USA).

Jaupart, C., Mareschal, J.-C., Rudnick, R. L., & Gao, S. (2004). *Treatise on geochemistry. The crust* (Vol. 3). Elsevier Pergamon.

Kesler, S. E., Gruber, P. W., Medina, P. A., Keoleian, G. A., Everson, M. P., & Wallington, T. J. (2012). Global lithium resources: Relative importance of pegmatite, brine and other deposits. *Ore Geology Reviews, 48,* 55–69.

Mohr, S. H., Mudd, G., & Giurco, D. (2012). Lithium resources and production: Critical assessment and global projections. *Minerals, 2*(1), 65–84.

Mudd, G. M. (2014). The future of Yellowcake: A global assessment of uranium resources and mining. *The Science of the Total Environment, 472,* 590–607.

Mudd, G. M., & Jowitt, S. M. (2014). A detailed assessment of global nickel resource trends and endowments. *Economic Geology, 109*(7), 1813–1841.

Mudd, G. M., Weng, Z., & Jowitt, S. M. (2013). A detailed assessment of global Cu resource trends and endowments. *Economic Geology, 108*(5), 1163–1183.

Palacios, J. L., Calvo, G., Valero, A., & Valero, A. (2018a). The cost of mineral depletion in Latin America: An exergoecology view. *Resources Policy, 59,* 117–124.

Palacios, J. L., Calvo, G., Valero, A., & Valero, A. (2018b). Exergoecology assessment of mineral exports from Latin America: Beyond a tonnage perspective. *Sustainability (Switzerland), 10*(723).

Sanchez-Garzoli, G. (2012). *Stopping irreparable harm: Acting on Colombia's Afro-Colombian and indigenous communities protection crisis.* Available at: https://reliefweb.int/sites/reliefweb.int/files/resources/04fcd8f818b16e1c31c4306ad74dfb70.pdf.

SIMCO. (2011). *Sistema de informacion minero colombiano, producción oficial de minerales en Colombia.* https://www1.upme.gov.co/simco/Paginas/home.aspx.

Sverdrup, H. U. (2016). Modelling global extraction, supply, price and depletion of the extractable geological resources with the LITHIUM model. *Resources, Conservation and Recycling, 114,* 112–129.

Sverdrup, H. U., & Ragnarsdottir, K.V. (2014). Natural resources in a planetary perspective. *Geochemical Perspectives, 3*(2).

Swain, B. (2017). Recovery and recycling of lithium: A review. *Separation and Purification Technology, 172,* 388–403.

Tilton, J. E., & Lagos, G. (2007). Assessing the long-run availability of copper. *Resources Policy, 32*(1–2), 19–23.

U.S. Department of Defense. (2015). *Strategic and critical materials. 2015 Report on stockpile requirements.*

UNEP. (2011). *Recycling rates of metals—A status report. A report of the working group of the global metal flows to the international resource panel.*

USGS. (2016). *Mineral Commodity Summaries 2016.* United States Geological Service. Available at: https://s3-us-west-2.amazonaws.com/prd-wret/assets/palladium/production/mineralpubs/mcs/mcs2016.pdf.

USGS. (2019). *Mineral Commodity Summaries 2019.* United States Geological Service. Available at: https://prd-wret.s3-us-west-2.amazonaws.com/assets/palladium/production/atoms/files/mcs2019_all.pdf.

Valero, A., & Valero, A. (2014). *Thanatia: The destiny of the earth's mineral resources: A thermodynamic cradle-to-cradle assessment.* World Scientific Publishing Company.

World Energy Council. (2016). *World energy resources 2016.* Available at: https://www.worldenergy.org/assets/images/imported/2016/10/World-Energy-Resources-Full-report-2016.10.03.pdf.

Wübbeke, J. (2013). Rare earth element s in China: Policies and narratives of reinventing an industry. *Resources Policy, 38,* 384–394.

Yaksic, A., & Tilton, J. E. (2009). Using the cumulative availability curve to assess the threat of mineral depletion: The case of lithium. *Resources Policy, 34*(4), 185–194.

Chapter 6
Material Limits of the Energy Transition

Abstract Currently, different scenarios are being created to analyse how to reduce greenhouse gas emissions and so limit the global temperature rise to 1.5 °C above pre-industrial levels. To reach this goal, we must reduce consumption and change the foundations of our energy system, and the most effective way seems to increase the use of renewable energy sources. Around 85% of world energy consumption is based on fossil fuels such as oil, coal and natural gas. That said, in recent decades, biomass consumption and the installed capacity of wind, hydro, solar, and other renewable energy sources have been continuously increasing. This chapter will explore this trend and the materials needed to build each clean technology. Combining this information with International Energy Agency scenarios, we will predict which elements could present possible supply shortages in the near future, putting at risk the very development of the energy transition.

Although globally, there is no tendency to reduce the extraction and consumption of raw materials, there has been in recent decades a growing interest in reducing our dependence on non-renewable energy sources such as fossil fuels, concerning the impact the burning of these fuels has on the atmosphere. The solution to a low carbon economy is to develop renewable energy vigorously. However, will this energy transition be as green as it is always presented to us? What influence will there be on the use of materials? We will discuss these issues in the present chapter.

6.1 The Paris Agreements and Climate Scenarios

Accelerating changes in the environment and industrial growth lead to the poles melting faster, causing sea levels to rise and large amounts of greenhouse gases—including carbon dioxide, nitrogen oxide and methane—to be released. The global average temperature for the period 2006–2015 was 0.87 °C, higher than the average for the period 1850–1900; this increase is fundamentally caused by human action. It is expected to increase by 0.2 °C per decade (IPCC, 2019).

© The Author(s), under exclusive license to Springer Nature Switzerland AG 2021   147
A. Valero et al., *The Material Limits of Energy Transition: Thanatia*,
https://doi.org/10.1007/978-3-030-78533-8_6

Consequently, the planet is experiencing extreme climate scenarios with greater frequency, including heatwaves, droughts and floods. Temperature variation is also expected to destroy large forest stands, generate mass extinctions of species and contribute to the expansion of certain diseases since transmitters that were previously only in tropical areas will migrate towards the poles as temperatures rise. In extreme scenarios, it is even possible for global warming to generate a change in the dynamics of Atlantic Gulf currents, altering the global ocean circulation (Warren, 2011). To avoid these consequences, we must reduce consumption and change the foundations of our energy system.

The Intergovernmental Panel on Climate Change (IPCC) is one of the organisations preparing technical assessment reports on climate change since 1988. The most recent IPCC report, commonly known as AR5, looks at the historical evolution of greenhouse gas concentration (GHG) emissions. Their results are conclusive: between 2000 and 2010, these emissions grew more than in the previous three decades (IPCC, 2014). This report also presents possible scenarios based on Representative Concentration Pathways (RCPs) that describe various situations for both GHG emissions and concentrations. Thus, the IPCC considers a scenario of substantial emission reductions (RCP 2.6), two intermediate scenarios (RCP 4.5 and RCP 6.0) and a scenario of high emissions (RCP 8.5). Various methods exist to limit the increase in temperature to 1.5 °C compared with pre-industrial levels, but these would require a substantial decrease in GHG emissions in the coming decades.

For example, the IPPC states that for RCP 2.6, the global temperature increase is unlikely to be less than 1.5 °C during the twenty-first century if the CO_2-eq concentration in parts per million was between 480 and 580. In contrast, it would be likely to reach a temperature below 2 °C. On the other hand, if the CO_2-eq concentration reached 720 ppm this century, a scenario included in the RCP 4.5, the probability of maintaining the temperature rise below 3 °C would be very low. However, it could be below an increase of 4 °C. In the worst-case scenario, where emissions are highest (RCP 8.5), it would be unlikely for the increase to be less than 4 °C with a CO_2-eq concentration of more than 1000 ppm.

In response to this evidence, during the 2015 Paris Climate Conference (COP21), a total of 195 countries set forth the first binding agreement to prevent climate change. They established a global action plan that limited the increase in temperature below 2 °C (aspiring not to exceed 1.5 °C) above pre-industrial levels (United Nations/Framework Convention on Climate Change, 2015). The conference proposed limiting greenhouse gas emissions as soon as possible, reducing them by up to 80% compared with 1990 levels. This agreement entered into force in November 2016 after being ratified by parties representing 55% of total greenhouse gas emissions.

In response to climate issues and as a consequence of COP21, the IPCC began in 2016 to report on the impact of a global temperature rise of 1.5 °C compared with pre-industrial levels. The report included guidance on how we should react to such an event by limiting greenhouse gas emissions (IPCC, 2019). The report establishes that if emissions continue at the current rate, this 1.5 °C rise will occur between 2030 and 2052.

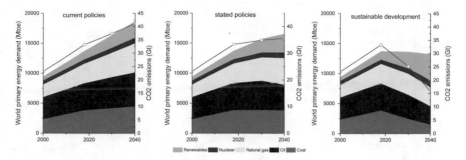

Fig. 6.1 Primary energy demand in the world by type of fuel and associated emissions for the three scenarios (according to the International Energy Agency (2019a))

This body is not the only one to present reports with projections and scenarios of how the situation could evolve based on GHG emissions. The International Energy Agency publishes an annual report on the energy sector's perspectives in the world (World Energy Outlook) where a series of future scenarios based on various hypotheses are established.

In the 2019 edition, three scenarios were presented. First, the current policies' scenario shows what would happen if the world continues as it is without implementing new environmental policies (Fig. 6.1). Second, the stated policies' scenario incorporates in its calculations existing political intentions and objectives, assuming that energy demand increases by 1% until 2040. Finally, the sustainable development scenario, which, through rapid and widespread changes in the energy system, allows for the total achievement of the sustainable energy objectives to keep the global temperature increase below 2 °C compared with pre-industrial levels (International Energy Agency, 2019a).

The International Energy Agency periodically publishes a report, *Energy Technology Perspectives* (*ETP*), analysing the technological advances that will shape sustainability as well as studies on future trends and advances that may produce a change in the energy sector (International Energy Agency, 2017). In the 2017 report, a total of three scenarios were established. The first, the reference scenario (RTS), only considers current commitments to limit emissions and carried out improvements in energy efficiency, which would result in an increase of 2.7 °C by 2100. In this first scenario, the policies being considered were current and a slight increase in CO_2 emissions globally is forecast. In the case of the other two scenarios, these emissions are considerably reduced. The second scenario (2DS) assumes a 70% reduction in CO_2 emissions in the energy sector from now until 2060, with a 50% probability of limiting the temperature increase to 2 °C. The third scenario (B2DS) is the most optimistic, where zero net emissions from the energy sector are reached by 2060, causing the temperature to rise by less than 2 °C.

Greenpeace also develops its own energy scenarios, being the best known those that appear in the 2015 report (Greenpeace, 2015). Specifically, among the various scenarios presented, the most striking is the Advance Energy [R]evolution (known as

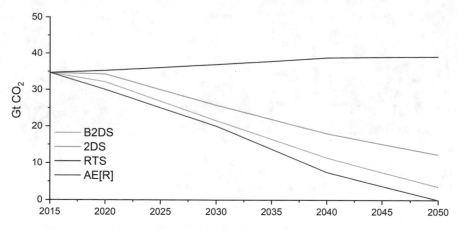

Fig. 6.2 Evolution of CO_2 emissions for different scenarios (International Energy Agency, 2017). RTS: reference scenario; 2DS: scenario with a 50% probability of limiting the temperature increase to 2 °C; B2DS: scenario with a temperature rise of less than 2 °C. AE [R]: 100% renewable scenario in 2050

AE[R]). This scenario assumes that, in 2050, the energy system will be completely decarbonised. One hundred per cent of the energy will come from renewable sources via energy-efficient measures, and renewable energies will be implemented as quickly as technically and economically possible.

Figure 6.2 shows a comparison between the projections of CO_2 emissions for each of these four scenarios from 2015 to 2050. As mentioned, the Greenpeace scenario is the most optimistic one, reaching zero emissions in 2050 based 100% on energy generation through renewable sources. The RTS scenario, which we would experience if we do nothing, is the most pessimistic, with emissions increasing compared with current levels.

It is also worth noting the reports prepared by organisations such as British Petroleum, which make forecasts on the global energy market in the coming years, the United States Energy Information Administration (EIA) or the World Energy Council (WEC) (BP, 2018; EIA, 2016; Greenpeace, 2015; World Energy Council, 2016). In addition, dynamic mathematical models exist, in line with Forrester's developments with his "World 3" model, which gave rise to the well-known book "The Limits to Growth" (Meadows et al., 1972). These models have the advantage of considering the interrelationships that exist between different variables and that combine technological issues in the energy sector with socio-economic and environmental issues. These models are beneficial not only for predicting what might happen in the future but also for analysing the cross-effects that adopting one scenario over another can have. These models also serve as models of energy demand, energy supply, impact, etc.

Models of this type include LEAP (Stockholm Environment Institute, 2005), TIMES (Loulou et al., 2005), WoLim (Capellán-Pérez et al., 2014), Witch (Bosetti et al., 2006), GCAM4 (Riahi et al., 2017), MESSAGE-GLOBIOM (Krey et al., 2016),

MEDEAS (Capellán-Pérez et al., 2020) and RETScreen (RETScreen International, 2009).

In general, these scenarios converge in the view that the electricity generation sector is the one with the most significant potential for reducing CO_2 emissions, making it perfectly feasible to eliminate them entirely by 2050 (European Commission, 2015). Emissions can also be substantially reduced in the transport sector, especially in private transport (European Commission, 2011). In this necessary transition period, renewable energies such as wind power, solar photovoltaic or electric vehicles will be of crucial importance in reducing emissions and transitioning to a decarbonised society. We will now look at the renewable energy sources that exist today and their evolution, before evaluating the materials they require (according to some of the projections seen above) and carrying out their corresponding exergy analyses.

6.2 Generation of Energy from Renewable Sources

So far, we have essentially focused our discussion on fossil fuels when talking about power generation: oil, coal and natural gas, which account for 85% of world energy consumption. However, other energy sources are considered renewable since they can regenerate on a human time scale. These include biomass, wind energy, hydroelectric, solar, energy from the oceans and geothermal energy. Each of these energy sources will be discussed in more detail below.

6.2.1 Biomass

Plants are naturally responsible for transforming energy from the Sun into chemical energy through photosynthesis. Some of this energy is stored in the form of organic matter, which can then be recovered by burning or transforming it into fuel.

The term biomass can include materials derived from plants such as wood, in the form of firewood or wood chips; agricultural crops for energy purposes; plantation residues and organic residues, such as manure from animal farms. All these products can be transformed into energy.

Estimates of the dry amount of all living matter on the planet vary, but it has an average value of around 2×10^{12} tons. Considering the specific energy of biomass as 17 MJ/kg, the useful energy content of dry biomass on land is approximately 810 Gtoe (Valero & Valero, 2014). Assuming this amount, it is not possible to exploit all this energy because this would imply deforestation and desertification.

The estimated theoretical biomass potential is about 92 TW, which implies an available useful energy capacity of 70 Gtoe per year (Johansson et al., 2004). Global biomass production varies greatly depending on the assumptions made. IPCC estimates show an energy potential of crude biomass of 10.4 Gtoe/year (14 TW) and liquid biofuels of 3.6 Gtoe/year (4.8 TW) (IPCC, 2001).

Global consumption of charcoal and firewood has been relatively stable for years, but the use of wood chips and wood pellets to generate electricity and residential heat has doubled in the past decade. It is expected to continue to increase in the future. In the European Union, biomass used for bioenergy continues to be one of the main sources of renewable energy, being used fundamentally for generating heat and cold (Scarlat et al., 2019).

From a global point of view, when biomass is burned, elements such as phosphorus, potassium and other microelements that the plant extracts are lost and must be replaced in the following harvest. Furthermore, these elements either produce slags that complicate combustion or contaminating NO_x. Therefore, biomass such as firewood and woody materials should be used in biomass plants, and the rest returned to the ground to close the cycle of the elements.

According to the International Energy Agency, in 2018, there were approximately 3,800 active biomass plants for electricity generation, with a capacity of 60 GW. For example, in Sandakan, Malaysia, there is an 11.5 MW plant that uses empty fruit bunches to generate electricity (Fig. 6.3). In addition, in Sangüesa (Navarra, Spain), there is another 30 MW biomass plant that has been operational since 2002.

Besides, there are liquid biofuels, including bioethanol and biodiesel. Biomass is currently the only renewable raw material that can be used to produce liquid fuel

Fig. 6.3 Photograph of the Seguntor bioenergy plant, located in Sandakan (Malaysia). Author: CEphoto, Uwe Aranas. Wikimedia Commons

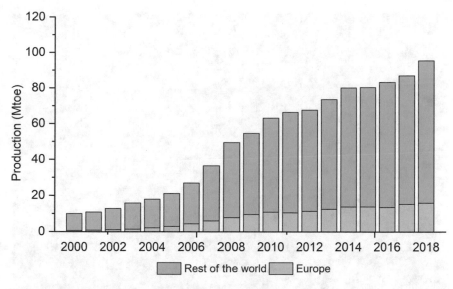

Fig. 6.4 Global production of biofuels. *Data source* BP

and that can be used as a substitute for gasoline for internal combustion engines. Commercial bioethanol is produced from crops based on sugar and starch, including sugarcane or sugar beet, with yields of up to 5,000 L per hectare. Biodiesel is primarily derived from vegetable oil or fat residues. At the top of the list of countries that produce the most biodiesel are the USA and Brazil, with a total of 36 and 20 Mtoe in 2018, respectively, according to the International Energy Agency.

Global production of biofuels is shown in Fig. 6.4, highlighting production in Europe and the rest of the world. America produced around 68% of biofuels, while Europe's production barely exceeded 16%. The clear growing trend in this type of production in recent decades is also observed, having multiplied almost tenfold from 2000 to 2018. Still, it should not be forgotten that biofuels compete with other land uses and that they are also associated with deforestation processes, as is the case in some areas where trees have been cut down to produce soybean oil (biodiesel) or to use grains (bioethanol) in cars.

6.2.2 Wind Power

The force of the wind has been exploited since ancient times, including sailing boats or windmills for grinding grain. However, it was not until the end of the nineteenth century that it began to be used to generate electricity through the use of wind turbines.

Its operation is relatively simple: the action of the wind moves the blades located at the top of the wind turbines and the rotational energy is transformed into electrical energy. There are different models of wind turbines depending on the location of the

Fig. 6.5 Middelgrunden wind farm, located at sea, about 3.5 km from Copenhagen (Denmark). This wind farm was built in 2000, currently, the most prominent marine farm in the world, boasting 20 turbines with a capacity of 40 MW (the distance between each wind turbine is about 180 m). Author: Kim Hansen, processing carried out by Richard Bartz and Kim Hansen. CC BY-SA 3.0. Wikimedia Commons

axis of rotation (horizontal or vertical), the type of interior generator, whether they are located on land (on-shore) or at sea (off-shore), etc. (Fig. 6.5).

The wind is an indirect expression of solar power, which is derived from pressure differences caused by uneven heating and cooling of the atmosphere and the rotation of the planet. The wind force varies considerably depending on the regions and the altitude; therefore, there is more wind in coastal areas due to the thermal differences between land and sea, with the same difference occurring in plains and mountainous areas. Installing a wind farm requires a preliminary wind study to determine the optimal arrangement of the turbines, their orientation, etc., to make the most of the energy. However, not all wind energy can be harnessed: in practice, a wind turbine can theoretically harness a maximum of 60% of the wind's energy. A turbine starts operating when the wind speed reaches between 3 and 4 m per second and reaches its maximum production at 13 or 14 m per second. However, if the wind speed is very high, the turbines are blocked for safety reasons, so it is not uncommon on a very windy day to see these devices out of operation.

Estimates of global wind power are very large (on the order of 10^{15} W), but most of that power is at high altitudes and cannot be recovered with equipment on the Earth's surface. The same situation occurs offshore. A technical potential of 72 TW

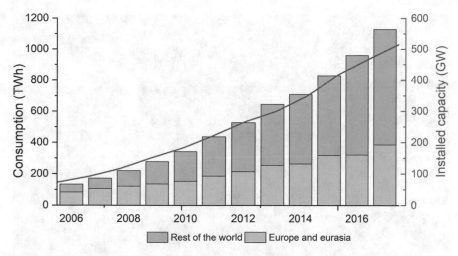

Fig. 6.6 Annual consumption of wind energy and installed capacity, including on- and offshore. Figure developed from BP and IRENA data

of global installed capacity, at 20% of the average capacity factor, would generate 126,000 TWh/year or around 14.5 TW(Valero & Valero, 2014).

In 2006, the existing global energy capacity was 73 GW (Fig. 6.6), with this figure increasing significantly to 594 GW in 2018 (IRENA, 2019a). With this growth in global wind power, this energy source already meets over 5% of the global demand for electricity. In addition, according to data from the Global Wind Energy Council (GWEC), wind energy is Europe's leading energy source generated natively, growing by 5.2% between 2017 and 2018 (BP, 2019).

6.2.3   Hydroelectric Power

Hydroelectric power uses energy from moving water to generate electricity. This type of energy was used in Ancient Greece to move wheat mills and make flour, and by the end of the nineteenth century, it had become a source of electricity generation.

About 23% of solar radiation constitutes the force that moves the hydrological cycle. The energy flow used for water evaporation is about 38,100 TW. This is transformed into potential energy from clouds (300 TW) and only a small part (5 TW) is transformed into potential energy from rivers. In addition, according to calculations of fresh water in the form of rain and snow, with an energy of 6 TW, the total energy available would total 11 TW (Valero & Valero, 2014).

Hydroelectric power can be used at different scales, from gigantic plants such as the Three Gorges Dam in China, with a capacity of 22.5 GW (Fig. 6.7) to plants of 100 W, 20 million times smaller. However, not all hydroelectric plants need a large dam to exist—sometimes it is enough to harness the energy of a watercourse.

Fig. 6.7 Photograph taken from the Three Gorges Dam International Space Station on the Yangtze River (China). This dam was completed in 2006 and has a capacity of 22.5 GW. The dam is about 2.3 km wide and about 185 m high. Author: NASA. CC BY 2.0. Wikimedia Commons

According to the International WaterPower & Dam Construction, the hydroelectric energy obtained from all exploited and exploitable places under technological limits and without considering environmental or economic restrictions is 1,800 GW.

In 2016, of all the electricity generated from renewable sources, hydroelectric power contributed 69%, that is, around 4,000 TWh, with China being the leading producing country followed by Canada and Brazil (IRENA, 2019a). Since 2000, installed power worldwide has been increasing at a rate of 3% per year (Fig. 6.8), concentrated mainly in developing countries, until 2018 when it reached 1,292 GW, including mixed plants (International Hydropower Association, 2019).

6.2.4 Solar Energy

The flux of solar radiation that warms the Earth and oceans is 43,200 TW, which is around 3000 times more energy than the world population needs annually: 14 TW in 2006 and 17.4 TW in 2015 (International Energy Agency, 2019a; Valero and Valero, 2014). The energy that the Sun provides us in 1 min is enough to supply all the

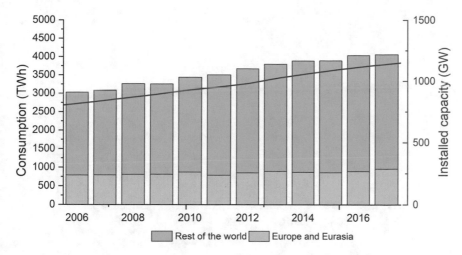

Fig. 6.8 Annual consumption of hydroelectric energy and installed capacity, including mixed installations. Figure developed from BP and IRENA data

world's energy needs in 1 year. Unfortunately, current technology is not sufficiently advanced to make use of this vast amount of energy.

Within solar energy, we can distinguish different technologies, among which are photovoltaic panels and solar thermal collectors, the former being the most developed globally. In the case of photovoltaic panels, electricity is obtained from solar radiation through photovoltaic cells, usually made from monocrystalline, polycrystalline or amorphous silicon. These boast a yield between 14 and 20%, although new technologies have recently been developed from the National Renewable Energy Laboratory and the Helmholtz Zentrum Berlin with records of over 45%. These cells belong to the so-called thin-film solar cells, composed of copper, indium, gallium and selenium (CIGS). They are used primarily for their high efficiency and for their lower cost compared with silicon. However, in terms of materials, they are composed of much more critical and scarce elements. Another type of well-known solar cell is one that uses a thin film of cadmium telluride (CdTe), responsible for absorbing and converting sunlight into electricity. The main advantage of this type of cell is that it is more cost-effective and has greater efficiency than other types. An efficiency of 22.5% was recently achieved (Martin, 2016), with its efficiency limit being over 30%. Still, this technology represents only about 5% of the global market. Perovskite, a material that can be easily synthesised and that could unseat the rest due to its low cost, ease of manufacture, flexibility, and high efficiency (Sivaram et al., 2015) should also be highlighted.

The accumulated installed capacity of solar photovoltaic has been experiencing a boom in recent years, reaching an almost exponential trend (Fig. 6.9), growing from 500 kW in 1977 to over 500 GW globally by the end of 2018 (Solar Power Europe, 2018; World Energy Council, 2016).

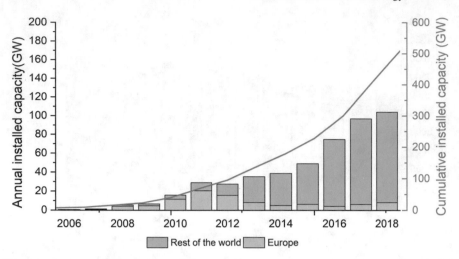

Fig. 6.9 Annual and accumulated installed capacity of solar photovoltaic in Europe and the rest of the world. Figure developed from Solar Power Europe data (2018)

Solar collectors, on the other hand, use solar energy to heat water. The installed solar thermal capacity in 2010 was 185 GWth and increased to 472 GWth by the end of 2017 (Fig. 6.10). China and Europe total 82% of the installed capacity worldwide (International Energy Agency, 2019b). In addition, installed capacity is expected to continue growing in the coming years.

Additionally, there are promising experiences in relation to Concentrated Solar Power (CSP). This type of solar energy uses mirrors or lenses to concentrate sunlight

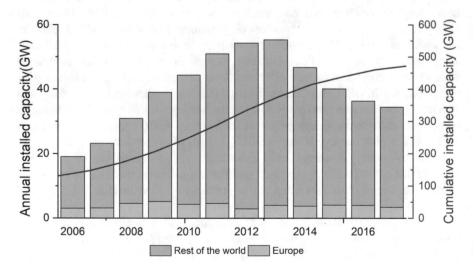

Fig. 6.10 Annual and cumulative installed capacity of solar thermal energy in Europe and the rest of the world. *Data sources* IEA

Fig. 6.11 Aerial view of the PS10 and PS20 central tower solar thermal plants (Seville, Spain). Author: Koza1983. CC BY 3.0. Wikimedia Commons

on a small surface, converting it to heat that drives a heat engine connected to an electricity generator. In Sanlúcar la Mayor, in Seville, there is a commercial solar thermal power station with a central tower and heliostat field, with a power of 11 MW, also known as PS10. Next to it, there is another called PS20, which has 20 MW of power (Fig. 6.11).

These experiences show that this technology can be an alternative way of producing clean electricity from the Sun. The installed capacity in 2018 of this type of technology was 5.5 GW (Ren21, 2019). Spain and the USA continue to be the leaders, however, there are a large number of projects that have been developed in countries such as the United Arab Emirates, India and China. Estimates of the technical potential for global electricity production with CSP vary widely. For example, the European Association for Solar Heating Electricity (ESTELA) and Greenpeace estimate a CSP capacity of 1,500 GWe by 2050.

6.2.5 Ocean Energy

The Sun is responsible for three effects that occur in the oceans: the oceanic thermal gradient, from which energy can potentially be extracted; thermohaline circulation, caused in part by the thermal gradient and which in turn causes the movement of huge volumes of water around the globe; and ocean waves, indirectly generated by the Sun through the wind.

The Sun heats the ocean surface, generating a thermal gradient that varies between 22 and 2 °C in the deep ocean. This temperature difference results in a potentially usable specific energy of about 800 J/kg of saltwater. Theoretically, this thermal gradient could be used to extract energy from the oceans. However, the small temperature difference involved makes the recovery of this thermal potential from the ocean impractical with current technology and taking into account the fact that no commercial plant currently exists. Considering that the mass of the oceans is equal to 1.37×10^{23} kg, this gives an absolute potentially usable energy of 1.13×10^8 Gtoe (Valero & Valero, 2014).

Another consequence of the thermal gradient of the oceans is the so-called thermohaline circulation. This global ocean circulation is due to density differences that depend on temperature and salinity. The transport of heat from the thermohaline circulation has been estimated at around 1,200 TW and 2,000 TW. The useful energy corresponding to this flow is about 100 TW transferred to the thermal gradient. Unfortunately, no technology can take advantage of this vast energy source.

Waves are another expression of solar energy. They are formed by the winds that blow over the ocean, and their energy content is several thousand times greater than the energy of the tides. The point at which wind transfers to ocean currents and waves is estimated at 60 TW, but internal friction reduces the useful energy of waves breaking at the surface to about 2.7 TW. There are different wave energy conversion schemes, but none has a large-scale use. The most commonly used equipment today are float wave energy converters, which are attached to the bottom by means of a submerged anchor. They also incorporate articulated mobile devices, a floating apparatus that follows the movement of the waves as if it was a snake, and tanks that have a hermetic upper part and a lower part connected to the sea and that take advantage of the oscillating movement of the waves.

An example of a plant that takes advantage of this wave energy can be found on the coasts of Portugal, near Peniche. It was installed in 2012, with three units of 100 kW of capacity each (Fig. 6.12).

In Europe, in 2018, 500 kW of plants that take advantage of wave energy were installed (Fig. 6.13), and plants that are currently in operation reach almost 12 MW in power (Ocean Energy Europe, 2018). Furthermore, it is expected that in the coming years, there will be more accelerated development of these technologies, particularly in countries such as Sweden, Ghana and the UK.

Tidal energy is the energy obtained by taking advantage of the tides, which result from the gravitational attraction exerted by the moon and to a lesser degree by the Sun. When the Earth rotates on its axis, two high and low tides occur daily everywhere in the world. Tidal heights are not uniform, rarely exceeding one meter in the deep ocean, but on the coast, they can reach 20 m. The movement of such amounts of water requires a large amount of energy estimated at 2.7 TW. This amounts to 0.85×10^{20} J each year (Valero & Valero, 2014).

Tidal energy constitutes an inexhaustible and non-polluting energy supply and regularly ensures power production year after year with less than 5% annual variation.

Fig. 6.12 Aerial view of the plant that harnesses the energy of the waves off the coast of Portugal. Author: AW-Energy Oy. CC BY 3.0. Wikimedia Commons

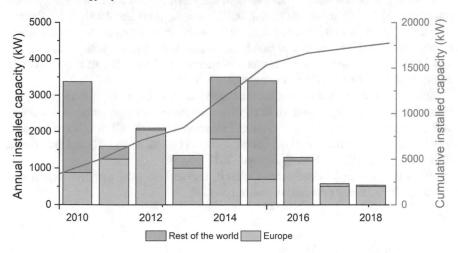

Fig. 6.13 Annual and cumulative installed capacity of wave energy in Europe and the rest of the world. *Data sources* Ocean Energy Europe

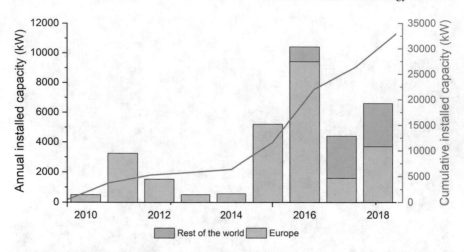

Fig. 6.14 Annual and accumulated installed capacity of tidal energy in Europe and the rest of the world. *Data source* Ocean Energy Europe

The high investment cost and the limited number of potential installation sites (about 20) are its main drawbacks.

There are very few tidal power plants in the world, with current installed power totalling 0.5 GW. Of these, the oldest, built between 1960 and 1966, is La Rance, with 240 MW of capacity, located in St. Malo, in Brittany, France. The largest plant, completed in 2011, is located in South Korea, on Lake Sihwa, with a capacity of 254 MW. In addition, several projects are planned in Scotland and France with a capacity of over 17 MW (World Energy Council, 2016).

Figure 6.14 shows the annual accumulated installed capacity of tidal energy in Europe and the rest of the world. Europe is where most new projects are installed, with the most relevant projects located in France and the UK. Of the almost 27 MW installed in Europe since 2010, 12 are currently in operation and the rest are still in the testing phase (Ocean Energy Europe, 2018).

According to the World Energy Council, the various tidal projects worldwide could reach an energy potential of about 166 GW.

6.2.6   Geothermal Energy

Geothermal energy is that which is stored or generated within our planet. The Earth's internal structure, separated into layers of different viscosities and temperatures, not only contributes to the transmission of internal heat but also implies a release of heat from the interior to the surface.

The heat flow in the terrestrial surface implies an increase in temperature with the depth of an average value of 2–4 °C per hundred meters. It is what is known as a normal geothermal gradient; this gradient, in anomalous zones, can exceed 20–30 °C every hundred meters of depth. These anomalous zones coincide with geologically active areas, such as regions with intense volcanic and seismic activity, ocean ridges, etc.

Geothermal reservoirs are classified according to their temperature, with low-temperature reservoirs that reach temperatures between 5 and 90 °C, used for direct heat exploitation in heating, fish farms, etc., up to high-temperature reservoirs, with temperatures higher than 150 °C, used for electricity generation. Today, dry hot rock deposits are also studied, where fluids are injected through artificial fracturing to establish a circuit with cold water injection and steam extraction used for electricity generation. Reservoirs with supercritical conditions are also being analysed, where the fluid is in an intermediate state between liquid and gas and can reach a temperature of 600 °C, although the technology required to exploit this type of resource is still very limited.

In general, the use of heat is produced by injecting fluids through a series of pipes in those areas of the subsoil where there are higher temperatures to be found; in these pipes, the fluid is heated; it then rises to the surface where, through a heat exchanger, its temperature is used for heating or cooling. This process can occur on a small scale, to heat or cool a building, but it also occurs on a large scale in geothermal power plants, where after the heat exchanger, a turbine transforms that thermal energy into electrical energy.

The Earth's continental crust can contribute between 5.8 and 6.9 TW to the global energy balance. The active provinces and continental margins currently represent 50% of the total volume of the crust. Extrapolating these values to the total surface of the Earth, it would imply global geothermal energy of 17.9 TW (Valero & Valero, 2014).

Geothermal energy is a renewable source of energy; however, its reserves represent only a small fraction of all geothermal heat. Furthermore, as is the case with tidal energy, geothermal energy may be significant locally. One example is Iceland, where 65% of primary energy comes from geothermal sources (Fig. 6.15). Still, the use of this type of energy is almost negligible on a global scale.

According to the latest report on renewable energy, global geothermal energy in 2018 was estimated at 630 PJ (175 TWh), half of which corresponding to electricity generation and the other half to direct use of energy (for heating or cooling) (Ren21, 2019). Furthermore, if we consider the total installed capacity of renewable energies, (including hydroelectric), geothermal energy barely reached 0.6% of the global amount in 2018.

6.2.7 Summary of Renewable Energy Sources

To summarise, Fig. 6.16 shows the evolution of the installed capacity of geothermal, hydroelectric, solar and wind energy, in addition to the corresponding part from biomass and the oceans, to produce electrical energy.

Fig. 6.15 Nesjavellir Geothermal Power Station (Iceland), the second-largest power plant in the country. Every year it produces about 120 MW of electrical energy in addition to hot water. Author: Gretar Ívarsson. Wikimedia Commons

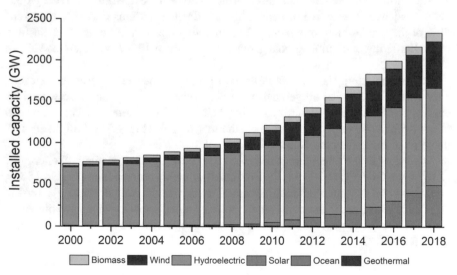

Fig. 6.16 Evolution of the installed capacity of the different renewable energy sources. *Data source* IRENA

Table 6.1 Available potential of each energy and installed capacity of 2018 (modified from Valero & Valero, 2014)

Type of energy	Available energy	Potentially usable energy	Installed capacity
Geothermal	17.9 TW	59–124 GWe	12.7 GW
Tidal	2.7 TW	166 GW	32,960 kW
Waves	3 TW	500 GW	17,745 kW
Oceanic thermal gradient	1.13×10^8 Gtep	–	–
Solar photovoltaic	43.2 PW	51 TW	509 GW
Solar heating	43.2 PW	630–4,700 GWe	472 GW
Hydroelectric	11 TW	1,800 GW	1,150 GW
Wind	72 TW	14.5 TW	594 GW
Biomass	92 TW	19–56 TW	60 GWe

As shown in the figure, the most relevant renewable energy source at a global level is hydroelectric, accounting for almost half of the installed capacity in 2018. Wind and solar energy compete for second and third place, with the remaining energies having less relevance since (aside from biomass) they cannot even be clearly seen in the figure due to the scale.

China, followed at a distance by the USA, is the country with the greatest installed capacity in renewable energy, specifically almost 700 GW, which is unsurprising since over 45% of renewable energy investments went to this country in 2017 (IRENA, 2019b). This means that China has reduced its dependence on the importing of fossil fuels and also the risks associated with possible interruptions in this supply. In addition, its technological experience in renewable energy has established it as a leading exporter of clean technologies, thus creating a commercial advantage compared to its counterparts.

Table 6.1 shows an approximation of the potentially available energy of each type of renewable energy, as well as data on the potentially usable energy and the installed capacity. These figures may vary depending on technological advances, which allow for more efficient use of resources and exploitation of those that are not yet economically viable today.

6.3 The Electric Vehicle

Transport is one of the sectors with the most significant potential in reducing greenhouse emissions by replacing conventional vehicles with an internal combustion engine for hybrid and plug-in vehicles, which can reduce emissions by 60% between now and 2050 (European Commission, 2011).

Electric vehicles use electricity as an energy source instead of gasoline or diesel. Within electric vehicles, we can distinguish two main groups: those that use the energy accumulated only in electric batteries (called BEVs) and those that use a combination of batteries with another source, such as gasoline, known as plug-in hybrid vehicles (PHEVs).

The origins of this type of vehicle go back to the nineteenth century when several inventors began to design different mobility types based on electricity. With the invention of rechargeable batteries, vehicles began to appear that used them as a source of energy. Gustave Trouvé presented the first vehicle that transported people at the International Electricity Exhibition in Paris (Fig. 6.17), and they were successfully tested on its streets in 1881 (Abarca Aguilar, 2019). That same year, another 21-battery powered electric four-wheeler, called Tilbury and designed by Charles Jeantaud, was also introduced.

After much progress, and with the twentieth century has already begun, combustion cars gained ground and it was not until a few decades ago that interest in a type of mobility based purely on electricity resurfaced. Today various models from different manufacturers are available, both 100% electric and plug-in hybrid types.

The evolution of electric vehicle sales globally can be seen in Fig. 6.18. In 2010, only a few thousand units were sold, but this figure has been increasing progressively, and quite rapidly, to this day. Specifically, in 2019, more than two million plug-in

Fig. 6.17 Engraving of the electric tricycle invented by Gustave Trouvé that appeared in the second volume of the book (Clerc, 1883)

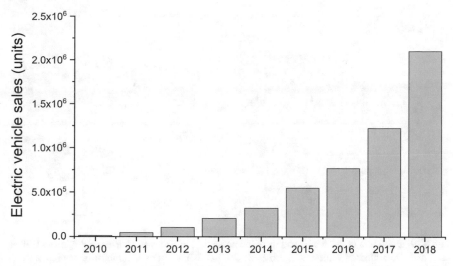

Fig. 6.18 Evolution of total sales of electric vehicles worldwide. *Data source* IEA

vehicle units were sold globally. China has seen the most units sold, closely followed by the United States (International Energy Agency, 2019c). In fact, sales of these green vehicles are expected to exceed those of conventional vehicles in the 2030s (ANFAC, 2014).

Given current policies that promote the acquisition of this type of vehicle, and the multiple benefits they offer the environment, the market sales trend is hardly surprising.

6.4 Cumulative Production of Low Carbon Technologies

We have seen that a societal transition towards a low carbon economy that allows us to maintain the planet's temperature rise below 2 °C implies a considerable increase in the use of renewable energy sources. To estimate the demand for materials, we need to select one of the scenarios mentioned in Sect. 6.1 to analyse the evolution of the associated demand for materials. It should be clarified that the figures presented here will not be more than mere estimates, based on various studies and agency reports, since it is impossible to predict exactly what will happen.

The future projections of each analysed renewable technology appear in Fig. 6.19. Only renewables with the greatest penetration have been considered because they will also be the ones that will use the most materials. In the case of the accumulated installed power, different bibliographic sources have been taken into account for each technology: wind energy, both onshore and offshore (EWEA, 2016; International Energy Agency, 2013), solar heating or thermoelectric energy (Greenpeace, 2016;

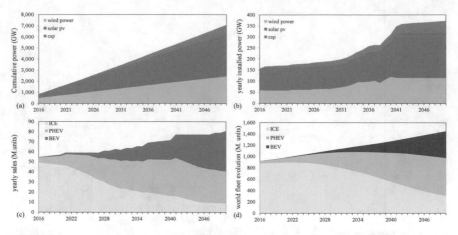

Fig. 6.19 Evolution of the annual and installed power of certain renewable technologies, as well as sales and a total fleet of vehicles (Valero et al., 2018a). ICEV: internal combustion vehicles; BEV: vehicles with electric batteries; PHEV: plug-in hybrid electric vehicles

International Energy Agency, 2014a) and photovoltaic solar energy (International Energy Agency, 2014b; Parrado et al., 2016).

Regarding the annual installed power of these technologies, the repowering effect has also been taken into account, given that the plants have to be updated or renewed from time to time, which is why a sudden change is observed in 2038, when all the renewable plants installed at the beginning of the century must be repowered.

Thus, in 2050, we could reach an annual installed power of wind, solar photovoltaic and solar heating of about 372 GW and total accumulated power of more than 7,000 GW: almost nine times more than the total accumulated power in 2016.

In the case of vehicles, annual sales and the total fleet of three models have been taken into account: internal combustion vehicles (ICEVs), electric battery vehicles (BEVs) and plug-in hybrid electric vehicles (PHEVs) (ANFAC, 2014; Dulac, 2013). In 2050, the fleet will consist of over 1.5 billion vehicles, of which ICEVs will represent a 21% share, PHEVs 45%, and, finally, BEVs, 33%. As we can see, sales of diesel and petrol vehicles will begin to decrease according to the estimates available from 2020 in favour of more sustainable vehicles.

In summary, sales and total installed capacity of renewable technologies will increase year after year, which, in theory, will allow us to halt the global increase in temperature. That said, all the technologies mentioned so far, our saviours, have certain limitations associated with the number of materials necessary to manufacture them. In the next section, we will look in greater detail at the material requirements of each of these technologies.

6.5 Materials for Low Carbon Technologies

Until now, we have described what type of technologies are going to play a fundamental role in the energy transition and also what the forecast of implementation of these technologies is according to various organisations. To achieve an optimal situation and to limit the unwanted temperature rise and mitigate the effects of climate change, renewable energy sources must not only be sustainable, they must also be competitive.

For a primary energy source to be competitive in the future, it must meet the following characteristics, according to Heinberg (2009): (1) it must be able to provide a substantial amount of energy and its contribution to countries must not be marginal; (2) the energy provided must be at least ten times greater than the energy required for placement from the cradle; (3) it must be acceptable from a social, geopolitical and environmental point of view; (4) it must be renewable. To these four principles, García-Olivares et al. (2012) add a fifth: it should not depend on scarce materials. This fifth point is precisely the most forgotten in practically all discourse in favour of the use of renewable energy. Along with the need for more research and development of renewable technologies, the infrastructure required per gigawatt produced is much greater than for conventional fossil fuel-based technologies and requires a larger amount of rare earths and raw materials.

For example, wind power, as we will now see, requires rare earths such as neodymium and dysprosium, used in the permanent magnets of electricity generators (Chakhmouradian et al., 2015; Elshkaki & Graedel, 2014). In the case of solar panels, significant amounts of silver are required for electrical connections, as well as other elements such as cadmium, tellurium and indium (Grandell et al., 2016). In the case of transport, conventional vehicles will be replaced by electric vehicles, both hybrid, and plug-in, and sales of the latter are expected to exceed those of internal combustion vehicles in early 2030 (ANFAC, 2014).

On the other hand, the sacrifice of using land for the implementation of renewable energies is not easy and causes environmental groups, despite being in favour of this type of alternative energy, to make their voices heard so that they are not installed near towns and residential areas, which can displace traditional agricultural or livestock activities, an effect known as *Not in my back yard* (NIMBY). This effect also applies to the opening of new mines, location of waste warehouses, etc. Simultaneously, the use of more materials to build wind turbines, solar panels, etc., implies greater primary extraction and greater energy consumption, which is also associated with a more significant impact on the environment. It is this impact, particularly in terms of material availability, that is not reflected in the energy models that so many agencies promote and develop, and where renewable energies play a leading role.

For this reason, we will now focus on the composition of these technologies to understand specifically what materials are needed today. To this aim, numerous studies of the materials used in each of them have been consulted, obtaining an average composition for wind turbines, electric vehicles of various types, solar panels, etc.

Table 6.2 Composition of two models of wind turbines, onshore and offshore (data in kg/MW) (Valero et al., 2018a)

	Model 1		Model 2	
	Onshore	Offshore	Onshore	Offshore
Al	840	840	560	560
Cu	2,700	11,500	7,000	15,800
Fe	172,100	292,100	112,670	232,670
Nd	60.92	60.92	182.75	182.75
Dy	4.86	4.86	14.58	14.58
Ni	111	111	111	111

In terms of wind turbines, both on land and at sea (Table 6.2), the key metals are iron and copper, used to manufacture the towers (iron) and all the required electrical and electronic components and cabling (copper). Other more minority elements are neodymium and dysprosium, belonging to the group of rare earths, used in the permanent magnets that the alternator incorporates, located in the nacelle, to transform the mechanical energy of rotation of the blades into electrical energy.

Among the renewable sources of energy, we find the hydroelectric plant. Despite their size and the fact that much of the materials needed to build the dams is concrete, they also require other materials. Table 6.3 summarises some of the most important, among which iron stands out. They also need two other elements, chromium and manganese, which are mainly used in alloys (Valero et al., 2018b).

As we have seen, the use of solar energy is based on different technologies, with different compositions of solar cells depending on the materials used. Its performance and cost will depend largely on this composition.

Table 6.4 shows the materials used in CSP plants, among which parabolic cylinders (PT) and central receiver systems (CRS) have been considered. Basic metals such as iron, aluminium, nickel and molybdenum predominate, as well as many other elements, appear in its composition. Specifically, nickel and molybdenum are used to make high-strength steels, copper for electrical wiring, and silver for high-performance solar crystals.

Regarding photovoltaic panels (Table 6.5), the most frequent is crystalline silicon (c-Si) where another of the main components is copper and to a lesser extent tin and silver. In the case of copper, indium, gallium and selenium (CIGS) solar cells, in addition to these four elements, we also see molybdenum and zinc. These have a lower cost than c-Si but elements such as indium and gallium are much scarcer

Table 6.3 Some of the materials needed in a hydroelectric power station (data in kg/MW) (Valero et al., 2018a)

Element	kg/MW
Cr	96,000
Fe	1,242,000
Mn	5,760

Table 6.4 List of materials used in concentrated solar power (CSP) (data in t/GW) (Valero et al., 2018a)

Element	PT	CRS
Ag	13	16
Al	740	23
Cr	2.2	3.7
Cu	3.2	1.4
Fe	650	393
Mn	2	5.7
Mo	200	56
Ni	940	1.8
Ti	25	0
V	2	2
Zn	650	1.4

PT: Parabolic cylinders; CRS: central receiver systems

Table 6.5 Average quantities of some elements used in different types of photovoltaic panels (data in t/GW) (Valero et al., 2018a)

Element	c-Si	CIGS	CdTe
Ag	133.0	–	–
Cd	6.1	1.8	65.2
Cu	4,177.5	19	42.8
Ga	0.1	4.9	–
In	4.5	23.2	15.9
Mg	53.5	–	–
Mo	–	94.3	100.5
Ni	1.1	–	–
Se	0.5	38.1	–
Si	6,326.5	–	–
Sn	520.0	–	6.6
Te	4.7	–	65.4
Zn	–	85.8	–

in comparison. In addition, the table includes the composition of cadmium telluride (CdTe) solar cells, also low in price and high efficiency.

Finally, in terms of solar energy, to be mentioned are those panels in charge of capturing sunlight to generate thermal or electrical energy and that is usually used to heat water in residential buildings and industries (Table 6.6). Regarding technology, there are both flat solar collectors and vacuum tubes, although worldwide vacuum tube technology predominates (International Energy Agency, 2019b).

Another renewable energy source, although much less widespread and whose use is usually restricted to the residential area, is geothermal energy (Table 6.7). The

Table 6.6 Average quantities of some elements used in solar thermal power (kg/MW) (Valero et al., 2018a)

Element	kg/MW	Element	kg/MW
Al	228.38	Mo	162.46
As	0.011	Ni	162.5
Cr	3,249.33	P	14.62
Cu	2,988.57	Pb	0.36
Fe	28,390.02	Si	1,615.29
K	37.3	Sn	0.04
Mg	149.86	Ti	21.2
Mn	324.94	Zn	4.72

Table 6.7 Average quantities of some elements used in heat pumps for geothermal energy (kg/MW) (Valero et al., 2018a)

Element	kg/MW
Al	6,790
Cr	200
Cu	2,440
Fe	14,900
Ni	240
Sn	3.6
Zn	110

materials correspond mainly to the composition of heat pumps. In this case, data have been taken from heat pumps of the GSHP type (Ground Source Heat Pump).

Table 6.8 shows the average composition of the two types of electric vehicles considered in this study: electric battery vehicles (BEVs) and plug-in hybrid electric vehicles (PHEVs). For comparison purposes, internal combustion vehicles (ICEVs) are also included. It should be noted that only a series of selected elements that are part of these vehicles appear in this table, not all of which are used for their production.

As can be seen, the new generation of vehicles will need many more materials, and more diverse, even more than those listed in this table, since in some cases, certain elements appear in such small quantities that it is impossible to know precisely how much is used. Two of these elements are lithium and cobalt, both used in battery models. This mobility change will also increase the demand for other elements, such as rare earths for permanent magnets, tantalum or indium for electronic components, etc. The use of other elements such as copper will also increase, which in theory is much more abundant. Its demand in the sector is expected in 2030 to be 250% times greater than the current demand (Wood Mackenzie, 2019).

After analysing the elements required to manufacture each of these low-carbon technologies, it is interesting to carry out a comparison with two other conventional technologies, such as gas and nuclear power plants (Table 6.9).

Table 6.8 Average quantities of some of the elements present in internal combustion vehicles (ICEVs), vehicles with electric batteries (BEVs) and plug-in hybrid electric vehicles (PHEVs)

Element	ICEV (diesel)	ICEV (gasoline)	PHEV	BEV
Ag	10.19	19.47	28.0	29.8
Al	61,103	78,343	141,370	200,000
Au	3.15	3.65	0.20	0.32
Ce	2.67	0.37	49.7	0.15
Co	9.72	8.06	2,712.4	9,330.6
Cr	5,041	5,566	6,510.0	6031.9
Cu	15,584	15,376	59,166	150,000
Dy	0.19	0.48	13.81	18.7
Eu	0	0.0001	0.23	0.23
Fe	701,095	653,524	806,144	746,945
Ga	0.27	0.27	0.81	1.12
Gd	0.0005	0.0005	0.17	0.17
Ge	0.0036	0.003	0.05	0.08
In	0.216	0.21	0.38	0.38
La	0.341	0.40	7.4	7.4
Li	22.06	4.63	2,242	7,709
Mn	4,333	4,211	5,968	5,530
Mo	240	188	260.0	260.0
Nb	154.2	145.6	426.3	426.3
Nd	23.7	18.8	553.8	749.3
Ni	1,590	2,993	16,049	55,724
Pb	12,527	11,532	9,750	9,750
Pd	1.99	1.84	0.94	0.0
Pr	0.066	0.08	51.5	98.0
Pt	3.79	0.13	5.51	0.0
Rh	0.12	0.09	0.01	0.0
Sm	0.21	0.33	2.3	3.2
Ta	4.65	6.53	10.8	10.8
Tb	0.01	0.02	13.6	26.9
V	92.81	86.62	852.6	790.0
Y	0.07	0.13	0.4	0.4
Yb	0.0003	0.0002	0.8	0.2

Data in grams per unit (Ortego et al., 2018; Valero et al., 2018a)

	Gas power plant (kg/MW)	Nuclear power station (kg/MW)
Al	750	200
Cr	0	2,190
Cu	750	1,470
Fe	5,500	58,904
Mn	0	75
Total	7,000	62,839

Table 6.9 Quantities of some of the elements present in gas and nuclear power plants (Valero et al., 2018a)

If we analyse these technologies one by one, we see that they all require a certain quantity and variety of elements, and this assuming that in the future, there will be no changes that alter their composition. Some of the elements they contain have already been seen in previous sections that are considered critical and strategic by various organisms, others are scarce in the crust, or very difficult to extract, or the mines from which they come are located in only a few countries globally.

Figure 6.20 shows a summary of the elements used in the different technologies analysed in this section. Hydroelectric technology requires the least variety of elements, in contrast to solar or electric vehicles. Besides, certain elements, such as iron, are present in all of these technologies, whereas others, such as germanium or tellurium, are required only in one specific type of technology.

Now that we understand the required materials for the development of the different energy technologies and the vehicle, we are able to exegetically evaluate the scenarios

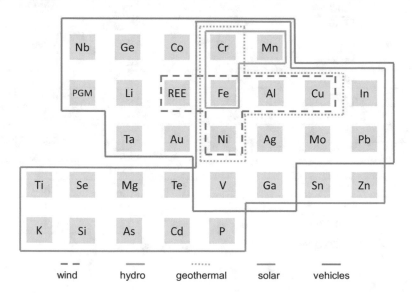

Fig. 6.20 Some of the materials required for the different types of renewable energy

of the energy transition considering the variation in energy and material flows, using the same units.

6.6   Exergy Flow Analysis of the Energy Transition Scenarios

The scenarios for the energy transition are usually represented by flows in the so-called "Sankey diagrams" in terms of energy and only refer to fossil fuels and other energy sources such as nuclear or, more recently, renewables. These Sankey diagrams represent, by means of arrows proportional to their size, the origin and destination of said energy flows. As we will see, if the unit of measurement is exergy as opposed to energy, we will not only be able to evaluate energy flows but also material flows through exergy replacement costs. We will thus be able to determine the influence and importance that mineral resources will have in the future as well as elucidating whether this transition can be genuinely qualified as renewable, given the intensity in the use of materials in renewable energy sources (Valero et al., 2018b).

For the purposes of this study, we have taken into account the renewable technologies indicated in Sect. 6.5, with their corresponding average compositions, assuming that they will not suffer variations until 2050: wind, solar photovoltaic, CSP, solar heating, geothermal, hydroelectric and the transport sector, with particular emphasis on electric vehicles (EVs) made up of vehicles with an electric battery (BEVs) and plug-in hybrid electric vehicles (PHEVs).

In the case of nuclear power plants, the required materials have been analysed based on the analysis of materials by prior technology, mainly in terms of steel, which also usually contain significant amounts of chromium and manganese. Furthermore, in the case of the rest of the conventional energy sources, it has been assumed that no new coal, natural gas or oil plants will be built, so new demand for materials of this type of energy will be equal to zero.

Based on the information on the composition of renewable technologies, information on the recycling of materials, the replacement rate of vehicles, repowering and future estimates of the increase in renewables, the demand for materials in the energy sector (both renewable and non-renewable) has been calculated from the present day to 2050 (Valero et al., 2018b). However, it is to be noted that there is a series of materials used in other sectors that must also be taken into account to obtain a more robust and comprehensive analysis. To perform the calculations, we have considered that the demand for materials from the non-energy sectors has an annual growth rate of 0.8%, equivalent to the expected growth of the population.

For the purposes of this study, we have employed the International Energy Agency's 2DS scenario (see Sect. 6.1), which stipulates that by 2100 emissions will have been limited such that there will be a 50% probability that the temperature will only rise by 2 °C with respect to pre-industrial levels. In this scenario, CO_2

emissions will reach a maximum in 2020, when they will begin to decrease until they reach a quarter of today's levels by 2060.

In Fig. 6.21, we can see a prediction of the energy sector in 2025 as well as the corresponding demand for materials for both energy and non-energy uses. All the values are represented in exergy terms in order to carry out a comparison between them (Valero et al., 2018b).

In the near future, we can see that society will still depend heavily on the consumption of natural gas, oil and coal, that is, on non-renewable sources of energy. The total energy generated is separated into two in each type of energy, which is destined for transformation and also is destined to directly generate electricity and heat.

In the case of renewables, although its use will increase a little, it still won't be enough to cover even a quarter of total demand. The sectors that will demand the most energy are industry, residential and transport.

In terms of minerals, the arrow's size is very significant, showing its particular relevance for society. A small part will go to the manufacture of renewable energy (394 Mtoe) and some also to the manufacture of non-renewable plants (114 Mtoe). Almost 700 Mtoe will go to the agriculture and fishing sector in the form of fertilisers and the rest will go to non-energy uses.

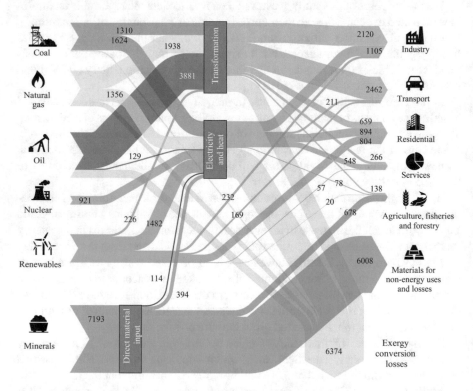

Fig. 6.21 Exergy flow based on the International Energy Agency's 2DS scenario for 2025, including mineral resources. Data in Mtoe (Valero et al., 2018b)

Figure 6.22 shows the prediction for 2050 (Valero et al., 2018b). At first glance, renewable energies will have a greater weight than fossil fuels and other traditional energy sources than in the previous prediction, given that the width of the total arrow is considerably greater.

In the case of transformation, fossil fuels continue to be the main source, oil in a greater proportion than others, but biofuels with a small fraction (584 Mtoe) also come into play. In terms of the destination sectors, industry and transport prevail over the rest.

As for the energy destined for the generation of electricity and heat, according to the IEA scenarios, in 2050, this will mainly come from renewable energy sources, although there will still be some dependence on nuclear energy and fossil fuels. This energy will be distributed in a similar way to all sectors, although industry continues to be the main driver.

Finally, if we look at the materials required, we can see that the size of the arrow has almost the same weight as renewable energies, that is, although in tons it does not seem significant, if compared in exergy terms, minerals acquire a very relevant weight in our economy. A small part of these materials will be destined for the production of

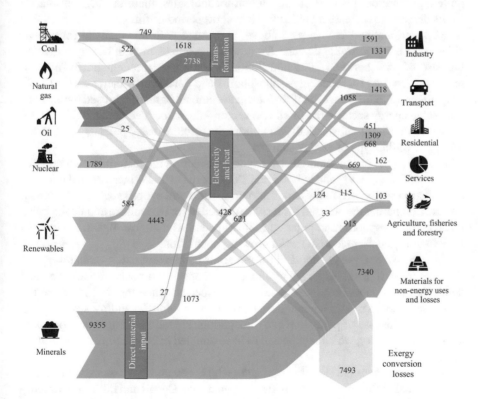

Fig. 6.22 Exergy flow based on the International Energy Agency's 2DS scenario for 2050, including mineral resources. Data in Mtoe (Valero et al., 2018b)

renewable technologies (1,073 Mtoe) and a small fraction for the production of non-renewable energy technologies (27 Mtoe). Agriculture is a sector that uses mineral matter directly, in the form of fertilisers. Given the vital importance of this sector, the arrow is also very representative in thickness compared to other sectors in this prediction (912 Mtoe).

It is true that most of the mineral extracted will be destined for non-energy purposes (7,340 Mtoe), and that, within this amount, a certain part will also correspond to losses, as in other sectors. However, most of these materials will either remain in the technosphere waiting to be recovered at some point or will dissipate irretrievably.

Comparing the 2025 (Fig. 6.21) prediction with that of 2050 (Fig. 6.22), there is a considerable decrease in total primary demand for fossil fuels, by 57, 31 and 27% for coal, oil and natural gas, respectively. In other words, in total, the primary demand for this type of most polluting energy source is expected to decrease almost a third in 25 years.

In that same period of time, the demand for energy from renewable sources is expected to increase by 131%, leading to an increase in nuclear energy by 94%. These hypotheses coincide with the general theory that our society must depend more and more on this type of low-carbon technologies, therefore it is reasonable for its demand to increase significantly in a short period of time.

However, and as we have seen above, associated with this demand for renewables is an increase in the demand for mineral resources that, in exergy terms: an increase of 35% between 2025 and 2050. For example, the demand for mineral resources required for bioenergy—phosphorus and potassium—increases by 172%, a significant amount. However, if we look at the amount of materials used for the non-renewable energy sector—the fossil fuel sector—the associated demand for mineral resources decreases substantially, an almost fivefold decrease, precisely due to the lower use of this type of energy sources.

Regarding the use of materials by sector, considering the materials used only for electric vehicles (BEVs) and hybrids (PHEVs), there is a 344% increase in demand in the transport sector, due to the expectation that sales will experience a strong annual uptick. Comparing the demand for fertilisers needed for agriculture and fishing, also made up of phosphorus and potassium, this increases by almost 40% since we will need to cultivate more land to feed a growing population.

We can also analyse the demand for materials by element and by technology for the 2DS scenario (Fig. 6.23). Aluminium, iron, copper and potassium are the most demanded materials. These results are to be expected since the first three are the most used metals in all technologies, from solar to wind, including steel used in vehicles. In the case of potassium and phosphorus, the demand only corresponds to the bioenergy sector, which is expected to have doubled in 2050 compared with 2025 levels.

In addition, there are six other elements whose demand increases sixfold in this period: cobalt, lithium, magnesium, titanium and zinc. Cobalt and lithium are mainly used in BEV and PHEV vehicles, while magnesium, titanium and zinc are used in CSP and solar heating.

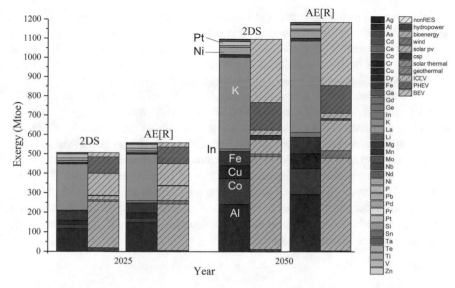

Fig. 6.23 Comparison of material requirements between the 2DS and AE [R] scenario for 2025 and 2050 by demand and by technology. Data in Mtoe (Valero et al., 2018b)

Some elements exist whose demand in the energy sector will decrease, such as chromium, lead and palladium. Chromium, although used in some green technologies, is also used in large quantities for ICEVs, and the same is true for palladium. Since sales of these types of vehicles are expected to decrease considerably, the associated demand for materials will decrease as well. It is important to note that this does not mean that the demand for these metals will decrease in a net way. This is because they are also used in other sectors and could even be used in future years in applications not yet known today, which could potentially affect their demand.

Information regarding the Greenpeace scenario AE[R], previously mentioned in Sect. 6.1, is also included in Fig. 6.23. In this scenario, the energy system should be fully decarbonised by 2050, with 100% of the world's electricity supply coming from renewable energy sources, with an installed production capacity of 23,600 GW (Greenpeace, 2015). This means that from 2020, emissions must decrease until they reach zero in 2050, with the total expected temperature increase being less than 2 °C.

In this Greenpeace scenario, the most optimistic scenario and the one in which the implementation of these renewable technologies occurs the fastest, this shift's fundamental pillars will be wind and solar energy, complemented by the remaining renewable sources. It also includes the use of hydrogen in the transport sector, in industry, for heating, etc., eventually completely replacing natural gas.

If we compare the demand for materials for the 2DS and AE[R] scenarios in 2025 and 2050, we can see that the Greenpeace scenario has a total demand of between 8 and 10% more materials (Fig. 6.23). In the case of some metals, such as cadmium, germanium, indium or tellurium, the demand associated with the energy sector shoots up to 200%. In others, such as magnesium, manganese, or zinc, demand decreases slightly. Regarding technologies, the most striking increase in materials occurs in geothermal, solar photovoltaic and wind energy.

In short, the new decarbonised scenarios indicate that the consumption of fossil fuels should be drastically reduced in favour of renewable energies, which will increase by 131% from 2025 to 2050. In exergy terms, the balance will be positive. That said, the transition will involve moving from a fossil fuel-dependent society to a highly mineral dependent society, some of them with significant supply risks, as we will discuss in the next section. According to data from the International Energy Agency, in the energy sector, the demand for materials in exergy terms will increase on average by around 30% in 2050. If other much more drastic scenarios in favour of renewables are taken into account, this demand may be even higher. This will undoubtedly increase the mining sector's pressure, which will again affect energy consumption since the average concentration of minerals in the mines will decrease as the deposits with higher concentrations are depleted, and it will take more energy to obtain the same amount of mineral.

6.7   Mineral Limits of the Energy Transition

The high dependence on minerals associated with the energy transition inevitably leads us to wonder if there are enough resources to satisfy the growing needs for raw materials. We must also ask ourselves whether the production rate will be able to adapt to demand at all times.

To answer these questions, we will first use the resource-based Hubbert peak data obtained in Sect. 5.2. Let us remember that, to find these peaks, we start from the resource data and the historical extraction of the different minerals. With this information, we can estimate the maximum extraction peak, that is, the moment in which the extraction stops increasing and begins to decrease when 50% of the available resources have already been exhausted. The limitations of this method have already been discussed since few minerals show an ideal bell-shaped extraction or the problem of elements that are extracted as by-products in the refining stages. However, this is a dynamic indicator that allows us to identify potential future issues, considering that resources do not increase and that the demand for raw materials increases exponentially.

Second, in order to know the future demand for materials from renewable technologies from now until 2050, we must know their current composition, assuming that there won't be great differences between current and future composition. This exercise has already been carried out in Sect. 6.5, so we will use the data from the aforementioned vehicle and renewable technology composition tables. It should be noted that, since each raw material is analysed one by one, the method presented here can be carried out by expressing the demand and reserves both in tons and in exergy replacement costs. The results will coincide.

Information on the demand for materials associated with renewable technologies should be combined with primary and secondary extractions. In addition to taking into account the future primary extraction obtained with the Hubbert curves, we must consider that a part of the material used to manufacture these technologies may come from recycling. This has been considered using the recycling rates of UNEP (UNEP, 2011) and using the following equation (Eq. 6.1):

$$d_{a_{gt}} = \left[\sum_{i-1}^{i-m} N \cdot M \cdot (1-r) \right] \tag{6.1}$$

where d_{a_gt} is the quantity of primary material demanded by the renewable technology being analysed during a specific year, N is the number of units manufactured each year, M is the amount of material used to make a unit and r is the amount of material derived from recycling.

Since the projections only go up to 2050 and some renewable technologies have a useful life that is inferior to the period being considered, the demand for repowering material and the renewal of the technologies and vehicles have been taken into account as follows (Eq. 6.2):

$$N = N_{ns} + N_{rm} N \tag{6.2}$$

where N_{ns} is the number of new units added to the global market and N_{rm} is the number of units manufactured to renovate the facilities that are dismantled.

In addition, it must be considered that all these renewable technologies are going to have to compete with many other sectors (industry, construction, chemistry, electronics…) in terms of the supply of materials. For example, today gallium is used not only in solar panels but also in integrated circuits, fundamental components of smartphones, military applications or wireless communication systems; gallium is also used in some types of automotive LED lights, in addition to alloys, batteries, permanent magnets, etc. It is thus important to take into account future demand not only for renewable technologies but also for other sectors.

Unfortunately, for the vast majority of sectors, the information available on material consumption is usually neither available nor publicly accessible. For this reason, this paper has assumed that the demand for materials from other sectors (d_{a_os}) will be constant until 2050, and equal to the difference between the total extraction of materials in 2015 $(P_a)_{2015}$ minus the demand for materials from renewable technologies $(d_{a\_gt})_{2015}$ (Eq. 6.3):

$$d_{a\_os} = (P_a)_{2015} - (d_{a_{gt}})_{2015} \tag{6.3}$$

Obviously, this is a very optimistic and conservative assumption, even more conservative than the one we made in the previous section, where we estimated that the consumption of materials for other sectors increased at a rate of 0.8%. That said, based on this assumption, we can estimate even more conservatively the possible supply risks and bottlenecks that may appear in the supply of materials in the renewable energy sector compared to other sectors.

Furthermore, the total material demand of a certain element a can be calculated in a specific year t (d_{a_T}) using the following equation (Eq. 6.4):

$$(d_{a\_T})_t = (d_{a\_gt})_t + (d_{a\_os})_t \tag{6.4}$$

Similarly, the total cumulative demand for an item will be D_{a_T} from 2018 to 2050, given by the following (Eq. 6.5):

$$D_{a\_T} = \sum_{t=2018}^{t=2050} \left[(d_{a\_T})_t\right] \tag{6.5}$$

Separating into percentages on the total accumulated demand of each renewable technology, we can see that the electric vehicle will demand a greater variety of elements (silver, aluminium, cerium, cobalt, copper, dysprosium, etc.) (Fig. 6.24).

Some elements, such as lanthanum, are only used in one of these renewable technologies, specifically in the electrical and electronic components of vehicles, but they are also used in many other sectors (catalysts, lighting, glass, ceramics...). Another example would be the case of silver: it is expected that between 2016 and 2050, almost 1 million tons will be needed to cover demand from all sectors. Around 34% of this demand will correspond to photovoltaic solar energy, 2.5% to electric

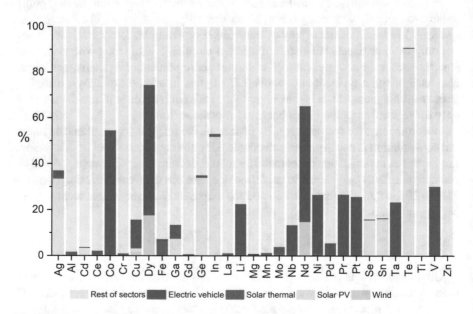

Fig. 6.24 Percentage of demand for materials by technology and by element with respect to the total accumulated demand from 2016 to 2050 (Valero et al., 2018a)

Table 6.10 List of defined risks and their corresponding definitions (Valero et al., 2018a)

Type of risk	Definition
Very high	Accumulated demand for 2016–2050 $\geq$ known resources $(d_{a\_T} \geq RES_{2015})$
High	Accumulated demand for 2016–2050 $\geq$ known reserves $(d_{a\_T}) \geq RSV_{2015})$
Medium	Annual demand $\geq$ annual primary extraction $(d_{a\_T})_t \geq (P_a)_t$

vehicles and 0.9% to solar thermal power, with the remainder corresponding to other sectors.

It is striking that a large part of the total accumulated demand for tellurium from 2016 to 2050 will be used in photovoltaic solar energy, or that more than half of dysprosium or neodymium will be used in wind energy and electrical vehicles.

Information on usage by sectors and total accumulated demand from 2016 to 2050 can be combined with information on available resources and reserves known today. In this way, we can determine the existence of possible supply shortages in some materials. To this end, three categories of risks have been defined, according to accumulated demand and total extraction: very high, high and medium (Table 6.10).

The most critical situation is when the accumulated demand for materials is greater than the amount of resources known today, since this would clearly jeopardise the supply of all sectors, including that of renewable technologies. The second category of risk is associated with a situation in which the accumulated demand exceeds the number of known reserves. To be noted is that reserves are much more variable than resources and that they are strongly influenced by the sale price of each item in question. The third category is associated with specific supply risks, that is, when the annual demand for a material exceeds the expected annual extraction, calculated using the Hubbert method—this method's flaws should be borne in mind.

Of all the analysed elements, only tellurium would present a very high risk. There are no tellurium mines as such: it is mined as a by-product of the refining of copper and lead–zinc minerals, so the available resource estimates are not accurate (Calvo and Valero, 2021). Thus, despite having the maximum risk, these data must be taken with caution.

If this accumulated demand is compared with the reserves, a much smaller, variable figure that depends on economic factors, the total accumulated demand would be greater than the reserves in the case of 13 elements: silver, cadmium, cobalt, chromium, copper, gallium, indium, lithium, manganese, nickel, platinum, zinc and of course telluride. For all these elements, the risk of supply would be high.

Finally, we found a total of 13 elements that could have a medium risk. Among them are tantalum, selenium, neodymium or molybdenum. Demand from other sectors that are not that of renewable energy has been considered constant over time, so a variation in this demand could lead to progress in the intersection between demand and extraction capacity.

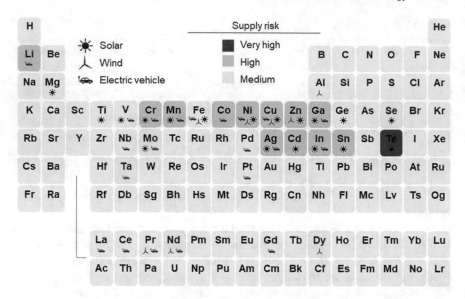

Fig. 6.25 Summary of the type of risk that each element presents and in which technology it is mostly used. Figure developed from data from Valero et al. (2018a)

Figure 6.25 summarises the types of risk of each element (very high, high and medium) as well as the technologies in which it is used (solar, wind, electric vehicle). As stated, the only element that presents a very high risk of supply is tellurium, marked in red, essential for the cadmium telluride (CdTe) solar cells mentioned above. Elements marked in orange present a high supply risk, whereas elements marked in yellow present a medium supply risk. Other elements are used in the analysed technologies but they do not present medium-term risks, such as titanium, vanadium or magnesium.

In total, there are thirteen elements that have a supply risk between high and very high, and that can therefore generate problems in the global supply of raw materials. These elements have been classified as critical in many studies, but not all at the same time. For example, in the case of the risk list prepared by the British Geological Survey (BGS), all are included except tellurium, although the risk classification varies according to each element (British Geological Survey, 2015). When compared to the list of critical elements drawn up by the European Commission in 2020, only four of these elements appear: cobalt, indium, gallium and lithium (European Commission, 2020). As discussed in Sect. 2.4, unlike in this analysis, traditional criticality studies do not focus on geological scarcity, but rather on socio-economic and geopolitical aspects.

In short, avoiding dependence on fossil fuels will mean accepting dependence on minerals. The demand for materials from renewable energies is going to be significant, not only in the short term but also in the medium and long term, potentially creating supply problems.

References

Abarca Aguilar, C. (2019). Motorización eléctrica y homologación de un coche deportivo mono-plaza para que pueda circular por ciudad. Available at: http://repositori.uji.es/xmlui/handle/10234/184587.

ANFAC. (2014). European motor vehicle parc. http://www.acea.be/uploads/statistic_documents/ACEA_PARC_2014_v3.pdf.

Bosetti, V., Carraro, C., Galeotti, M., Massetti, E., & Tavoni, M. (2006). WITCH: A world induced technical change hybrid model. *Energy Journal, 27*, 13–38.

BP. (2019). British petroleum. Statistical Review of World Energy. Available at: https://www.bp.com/content/dam/bp/business-sites/en/global/corporate/pdfs/energyeconomics/statistical-review/bp-stats-review-2019-full-report.pdf.

BP. (2018). *Energy outlook* (2018 ed.). Available at: https://www.bp.com/content/dam/bp/business-sites/en/global/corporate/pdfs/energy-economics/energy-outlook/bp-energy-outlook-2018.pdf.

British Geological Survey. (2015). Risk list 2015. http://www.bgs.ac.uk/mineralsuk/statistics/risklist.html.

Calvo, G., & Valero, A. (2021). Strategic mineral resources: Availability and future estimations for the renewable energy sector, Environmental Development, 100640, ISSN 2211-4645, https://doi.org/10.1016/j.envdev.2021.100640.

Capellán-Pérez, I., de Blas, I., Nieto, J., de Castro, C., Miguel, L. J., Carpintero, Ó., Mediavilla, M., Lobejón, L. F., Ferreras-Alonso, N., Rodrigo, P., Frechoso, F., & Álvarez-Antelo, D. (2020). MEDEAS: A new modeling framework integrating global biophysical and socioeconomic constraints. *Energy & Environmental Science, 13*, 986–1017. https://doi.org/10.1039/C9EE02627D.

Capellán-Pérez, I., Mediavilla, M., de Castro, C., & Miguel, L. J. (2014). World limits model (WoLiM) 1.0. Model documentation. Technical report.

Chakhmouradian, A. R., Smith, M. P., Kynicky, J. (2015). From "strategic" tungsten to "green" neodymium: A century of critical metals at a glance. *Ore Geology Reviews*. https://doi.org/10.1016/j.oregeorev.2014.06.008.

Clerc, A. (1883). Physique et chimie populaires.

Dulac, J. (2013). Global transport outlook to 2050. International Energy Agency. https://www.iea.org/media/workshops/2013/egrdmobility/DULAC_23052013.pdf.

EIA. (2016). International energy outlook 2016. U.S. Energy Information Administration.

Elshkaki, A., & Graedel, T. E. (2014). Dysprosium, the balance problem, and wind power technology. *Applied Energy, 136*, 548–559.

European Commission. (2020). Study on the EU's list of critical raw materials (2020). Final report. https://doi.org/10.2873/904613.

European Commission. (2015). 2050 low-carbon economy. Available at: http://ec.europa.eu/clima/policies/strategies/2050/index_en.htm.

European Commission. (2011). White paper. Roadmap to a single European transport area—Towards a competitive and resource efficient transport system. COM(2011) 144 final.

EWEA. (2016). The European offshore wind industry—Key trends and statistics 2015.

García-Olivares, A., Ballabrera-Poy, J., García-Ladona, E., & Turiel, A. (2012). A global renewable mix with proven technologies and common materials. *Energy Policy* 561–574.

Grandell, L., Lehtilä, A., Kivinen, M., Koljonen, T., Kihlman, S., & Lauri, L. S. (2016). Role of critical metals in the future markets of clean energy technologies. *Renewable Energy, 95*, 53–62. https://doi.org/10.1016/j.renene.2016.03.102.

Greenpeace. (2016). Solar thermal electricity. Global outlook 2016. Available at: https://www.greenpeace.org/archive-international/Global/international/publications/climate/2016/Solar-Thermal-Electricity-Global-Outlook-2016.pdf.

Greenpeace. (2015). Energy [R]evolution 2015. https://www.greenpeace.org/archive-international/en/campaigns/climate-change/energyrevolution/.

Heinberg, R. (2009). Searching for a miracle. Net energy limits & the fate of industrial society. Available at: https://www.postcarbon.org/publications/searching-for-a-miracle/.

International Energy Agency. (2019a). World energy outlook 2019.

International Energy Agency. (2019b). Solar heating & cooling programme. 2018 annual report.

International Energy Agency. (2019c). Global EV outlook 2019. Scaling up the transition to electric mobility. Available at: https://www.iea.org/reports/global-ev-outlook-2019.

International Energy Agency. (2017). Energy technology perspectives 2017. Catalysing energy technology transformations. International Energy Agency.

International Energy Agency. (2014a). Technology roadmap. Solar thermal electricity. Available at: https://www.iea.org/publications/freepublications/publication/technologyroadmapsolartherm alelectricity_2014edition.pdf.

International Energy Agency. (2014b). Technology roadmap. Solar photovoltaic energy. Available at: https://www.iea.org/publications/freepublications/publication/TechnologyRoadmapSola rPhotovoltaicEnergy_2014edition.pdf.

International Energy Agency. (2013). Technology roadmap. Wind Energy. Available at: https://www.iea.org/publications/freepublications/publication/Wind_2013_Roadmap.pdf.

International Hydropower Association. (2019). Hydropower status report. Sector trends and insights.

IPCC. (2019). Global warming of 1.5°C. Available at: https://www.ipcc.ch/sr15/.

IPCC. (2014). Climate change 2014: Synthesis report. In Core Writing Team, R.K. Pachauri and L.A. Meyer (Eds.), Contribution of Working Groups I, II and III to the fifth assessment report of the intergovernmental panel on climate change (p 151). IPCC.

IPCC. (2001). Climate change 2001 (Mitigation). Contribution of Working Group III to the third assessment report of the intergovernmental panel on climate change. In Technological and economic potential of greenhouse gas emissions reduction. Cambridge University Press.

IRENA. (2019a). Renewable statistics. Available at: https://www.irena.org/-/media/Files/IRENA/Agency/Publication/2019/Jul/IRENA_Renewable_energy_statistics_2019.pdf.

IRENA. (2019b). A new world. The geopolitics of the energy transformation. Global commission on the geopolitics of energy transformation. Available at: https://irena.org/-/media/Files/IRENA/Agency/Publication/2019/Jan/Global_commission_geopolitics_new_world_2019.pdf.

Johansson, T. B., McCormick, K., Neij, L., Turkenburg, W. (2004). The potentials of renewable energy. In: *Proceedings for the International Conference for Renewable Energies*, 1 to 4 June 2004, Bonn, Germany, pp. 40.

Krey, V., Havlik, P., Fricko, O., Zilliacus, J., Gidden, M., Strubegger, M., Kartasasmita, G., Ermolieva, T., Forsell, N., Gusti, M., Johnson, N., Kindermann, G., Kolp, P., McCollum, D. L., Pachauri, S., Rao, S., Rogelj, J., Valin, H., Obersteiner, M., Riahi, K. (2016). MESSAGE-GLOBIOM 1.0 Documentation. International Institute for Applied Systems Analysis (IIASA), International Institute for Applied System Analysis (IIASA), Schlossplatz 1, 2361 Laxenburg, Austria.

Loulou, R., Remne, U., Kanudia, A., Lehtila, A., Goldstein, G., 2005. Documentation for the TIMES model—PART I, pp. 1–78.

Martin, R. (2016). Células solares de teluro de cadmio logran un nuevo récord de eficiencia. Available at: https://www.technologyreview.es//s/5648/celulas-solares-de-teluro-de-cadmiolog ran-un-nuevo-record-de-eficiencia.

Meadows, D. H., Meadows, D. L., Randers, J., Behrens, W. W. (1972). *The limits to growth*. Universe Books.

Ocean Energy Europe. (2018). Ocean Energy. Key trends and statistics 2018.

Ortego, A., Valero, A., & Valero, A., Restrepo, E. (2018). Vehicles and critical raw materials. A sustainability assessment using thermodynamic rarity. Journal of Industrial Ecology. https://doi.org/10.1111/jiec.12737.

Parrado, C., Girard, A., Simon, F., & Fuentealba, E. (2016). 2050 LCOE (levelized cost of energy) projection for a hybrid PV (photovoltaic)-CSP (concentrated solar power) plant in the Atacama Desert, Chile. *Energy, 94*, 422–430. https://doi.org/10.1016/j.energy.2015.11.015.

Ren21. (2019). Renewables 2019. Global status report. París. Available at: https://www.ren21.net/wp-content/uploads/2019/05/gsr_2019_full_report_en.pdf.

RETScreen International. (2009). RETSCreen Model.

Riahi, K., van Vuuren, D. P., Kriegler, E., Edmonds, J., O'Neill, B. C., Fujimori, S., Bauer, N., Calvin, K., Dellink, R., Fricko, O., Lutz, W., Popp, A., Cuaresma, J. C., KC, S., Leimbach, M., Jiang, L., Kram, T., Rao, S., Emmerling, J., Ebi, K., Hasegawa, T., Havlik, P., Humpenöder, F., Da Silva, L. A., Smith, S., Stehfest, E., Bosetti, V., Eom, J., Gernaat, D., Masui, T., Rogelj, J., Strefler, J., Drouet, L., Krey, V., Luderer, G., Harmsen, M., Takahashi, K., Baumstark, L., Doelman, J. C., Kainuma, M., Klimont, Z., Marangoni, G., Lotze-Campen, H., Obersteiner, M., Tabeau, A., Tavoni, M. (2017). The shared socioeconomic pathways and their energy, land use, and greenhouse gas emissions implications: An overview. *Global Environmental Change, 42*, 153–168. https://doi.org/10.1016/j.gloenvcha.2016.05.009.

Scarlat, N., Dallemand, J., Taylor, N., & Banja, M. (2019). Brief on biomass for energy in the European Union, Sanchez Lopez, J. and Avraamides, M. editor(s), Publications Office of the European Union, Luxembourg, 2019, ISBN 978-92-79-77234-4, https://doi.org/10.2760/49052, JRC109354.

Sivaram, V., Stranks, S. D., Snaith, H. J. (2015). Placas solares de perovskita. *Investig. Cienc. 468*, 36–51.

Solar Power Europe. (2018). Global market outlook for solar power, 2019–2023. Available at: https://www.solarpowereurope.org/wp-content/uploads/2019/07/SolarPower-Europe_Global-Market-Outlook-2019-2023.pdf.

Stockholm Environment Institute. (2005). LEAP. Long-range Energy Alternatives Planning System. User guide for LEAP 2005.

UNEP. (2011). Recycling rates of metals—A status report. A report of the working group of the global metal flows to the International Resource Panel. Available at: https://www.resourcepanel.org/reports/recycling-rates-metals.

United Nations/Framework Convention on Climate Change. (2015). Paris Agreement. 21st Conf. Parties 3. FCCC/CP/2015/L.9.

Valero, A., Valero, A., Calvo, G., & Ortego, A. (2018a). Material bottlenecks in the future development of green technologies. *Renewable and Sustainable Energy Reviews, 93*, 178–200.

Valero, A., Valero, A., Calvo, G., Ortego, A., Ascaso, S., & Palacios, J. L. (2018b). Global material requirements for the energy transition. An exergy flow analysis of decarbonisation pathways. *Energy, 159*, 1175–1184.

Valero, A., & Valero, A. (2014). *Thanatia: The destiny of the Earth's mineral resources: A thermodynamic cradle-to-cradle assessment.* World Scientific Publishing Company.

Warren, R. (2011). The role of interactions in a world implementing change adaptation and mitigation solutions to climate. *Philosophical Transactions of the Royal Society, 369*, 217–241.

Wood Mackenzie. (2019). Global copper long-term outlook.

World Energy Council. (2016). World Energy Scenario 2016, World Energy Council (WEC). London. UK. 9780946121571.

Chapter 7
The Hidden Cost of Technologies

Abstract The hidden costs of technologies, considering the physical quality of the elements of which they are composed, will be analysed through a thermodynamic approach. First, the thermodynamic rarity of electrical and electronic devices is calculated and compared with the wastes this sector generates. Additionally, this same analysis is carried out with vehicles, considering the metals used in conventional and hybrid and electric cars. A particular emphasis is done on electric vehicle batteries. According to their rarity, a total of 31 vehicle components were identified as critical, being the engine and the gearbox the most critical ones due to their metal composition and low rate recovery of certain metals. Currently, the reuse and recycling rate of an end-of-life vehicle must be at least 85% by weight. Yet, we are far from effectively recovering all the elements that are used in vehicles. As we will see, out of every four vehicles entering the end of life stage, we only recover three in terms of thermodynamic rarity. This implies that some of the most critical elements are lost or dispersed, shifting supply pressure to the mining sector.

We're relying more and more on a plethora of electrical and electronic devices that seem to simplify our lives. Sales of this type of technology have only increased in recent years. In fact, in 2018, over 1.55 billion smartphones and almost 140 million electronic tablets were sold (Lui, 2019). These devices are composed of many base elements such as iron, copper or aluminium, as well as other minor metals that are scarce and whose extraction requires vast amounts of energy. If we consider that in Europe alone, a total of 10 million smartphones are replaced every month, it is not surprising that some of the elements that compose them will likely present a significant supply risk in the future.

The increase in sales and the constant renewal of these devices led to an increase in generated waste. Unfortunately, in many cases, these devices are not appropriately managed when discarded. For example, a UN report estimated the generation of waste composed of electrical and electronic devices at 50 million tons in 2019 alone. This figure includes not only smartphones and computers but also televisions, toasters, vacuum cleaners and large household appliances such as washing machines or refrigerators. To put this into perspective, 50 million tons is the equivalent in weight to almost 4,500 Eiffel Towers (World Economic Forum, 2019). Of this amount, only

20% were recycled through the corresponding management systems (Ku & Hung, 2014). Given that the expected growth in the sale of these devices is 3% per year, we could be looking at a total of 120 million tons of this type of waste generated by 2050 (Nijman, 2019).

In one ton of smartphones, we find several hundred grams of gold, meaning that the concentration of this element is up to 80 times greater than in current mines. Smartphones could therefore be a very interesting source of recovery. This happens with gold and with many other elements that appear in the composition of a smartphone, a tablet or a computer. What then is the physical value of these devices? In this chapter, we will attempt to answer this question using thermodynamic rarity, considering the physical quality of the elements of which they are composed. We can already answer questions such as whether an incandescent bulb is more efficient than a fluorescent one. Still, the answer may not be so evident if our evaluation considers manufacturing materials and their recycling pathway in addition to mere energy consumption. We will also carry out this same analysis for cars since they have arguably become the everyday consumer product with the highest concentration of known electrical and electronic components.

7.1   Thermodynamic Rarity of Electrical and Electronic Devices

Just as we did in the previous chapter for renewable energy, it is possible to evaluate any object's physical value by considering its elemental composition. Through exergy replacement costs, we consider the physical scarcity of the resources used for its production, whereas through thermodynamic rarity, we also take into account the energy required to extract and refine each element.

Let us take, for example, the composition of a smartphone, specifically the 16 GB iPhone 6. Approximately 39% of its weight is composed of iron and aluminium, with other metals such as copper, cobalt, chromium and nickel, which together total 17% of its mass (Merchant, 2017). If we take a closer look, we also find 0.014 g of gold, 0.01 g of gallium and 0.02 g of tantalum. There are many other elements too, such as rare earths, that appear in such small quantities that it is impossible to quantify them in an analysis of this type. The market price of metals contained in this phone barely reaches the dollar, with over half corresponding to gold. However, this market price fails to reflect the device's true value (Table 7.1).

Knowing the composition of any device A in terms of its elements i and their corresponding weights m_i in grams, we can evaluate the total thermodynamic rarity, Rarity(A) in kJ/g, as follows (Eq. 7.1):

$$\text{Rarity}(A) = \sum_{i=2}^{n} m_i \cdot \text{rarity}_i \qquad (7.1)$$

Table 7.1 Composition of an iPhone 6 (16 GB) and market price of each item (Merchant, 2017)

Element	Grams	Average price of a gram ($)	Price of contained metal ($)
Aluminium	31.14	0.0018	0.055
Arsenic	0.01	0.0022	–
Sulphur	0.44	0.0001	–
Bismuth	0.02	0.0110	0.0002
Calcium	0.44	0.0044	0.002
Carbon	19.85	0.0022	–
Chlorine	0.01	0.0011	–
Cobalt	6.59	0.0396	0.261
Copper	7.84	0.0059	0.047
Chrome	4.94	0.0020	0.010
Tin	0.66	0.0198	0.013
Phosphorus	0.03	0.0001	–
Gallium	0.01	0.3304	0.003
Hydrogen	5.52	–	–
Iron	18.63	0.0001	0.002
Lithium	0.87	0.0198	0.017
Magnesium	0.65	0.0099	0.006
Manganese	0.29	0.0077	0.002
Molybdenum	0.02	0.0176	–
Nickel	2.72	0.0099	0.027
Gold	0.014	40,00	0.56
Oxygen	18.71	–	–
Lead	0.04	0.0020	–
Potassium	0.33	0.0003	–
Silica	8.14	0.0001	0.001
Tantalum	0.02	0.1322	0.003
Titanium	0.3	0.0198	0.006
Tungsten	0.02	0.2203	0.004
Vanadium	0.04	0.0991	0.004
Zinc	0.69	0.0028	0.002
Total	129		1.03

Thus, the thermodynamic rarity of the iPhone 6 is approximately 236 MJ, a significant figure when compared with older models, which were around 184 MJ (Stanek et al., 2017). Considering the fact that almost 75 million units of this model have been sold in the world, of which very few are recycled effectively, the lost mineral wealth associated with this particular model would correspond to over 400,000 tons of oil equivalent, measured in thermodynamic rarity.

If in addition to smartphones, we include other small- and medium-sized devices, it is possible to measure the thermodynamic rarity concerning the sales of each one of them. A specific case study was carried out for 2010 sales of a total of 30 electrical and electronic devices in Germany, including LED lamps, DVD players, televisions of different models, headphones, smartphones, tablets, shavers, coffee makers, etc. (Valero et al., 2016). Specifically, ten metals present in all these devices were analysed: cobalt, gallium, gold, indium, palladium, rare earths, silver, tantalum, tin and yttrium.

The result of this analysis is shown in Fig. 7.1. The devices shown in the highest portion have greater thermodynamic rarity, that is, their components are more critical from a point of view of the properties of the materials. Also, devices on the right side of the graph are those with the most units sold: LEDs, laptop and desktop computers, smartphones and devices with GPS.

As expected, traditional devices with more basic composition, such as coffee machines, watches or razors, present considerably lower values of thermodynamic rarity. Furthermore, technologies that may seem more efficient are not necessarily more "sustainable" from the point of view of the materials they are composed of.

Let's look at an example related to lighting. Let us suppose we want to light a 20m^2 living room, for which we need a series of light bulbs, and we are interested in the most sustainable solution in terms of both consumption of materials and energy

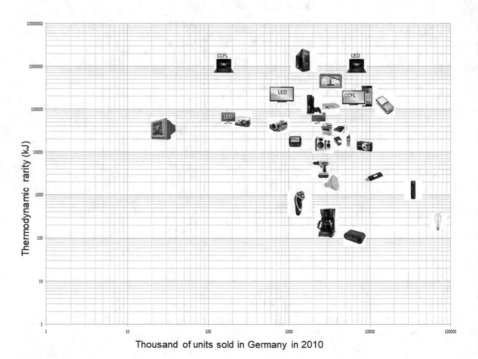

Fig. 7.1 Thermodynamic rarity of sales of a number of electrical and electronic devices in Germany in 2010 (Valero et al., 2016)

Table 7.2 Comparison between different types of light bulbs

	Units	Power consumption of 1000 h of operation	Metallic composition
Incandescent lamps	5 × 100 W	500 kWh	Tungsten filament, aluminium
Fluorescent lamps	5 × 20 W	100 kWh	Rare earth (1–15%), mercury, copper, iron
LEDs	6 × 10 W	60 kWh	Aluminium, gallium, indium, gold, silver, zinc, rare earth

consumption (Table 7.2). If we add the energy necessary to light the room throughout the life cycle of the bulbs, considering their thermodynamic rarity, the incandescent lamp is surprisingly more sustainable than the fluorescent one, as it requires far fewer materials. LED lights are the most efficient solution, requiring ten times less energy to operate, although they contain the most critical materials.

As it can be seen, the content of critical materials is a crucial component that must be taken into consideration for a more complete evaluation of product sustainability. In our previous calculations, we have assumed that these materials are not recycled. As we will see in the next chapter, the recycling rate for critical materials is extremely low. If the elements present in the products were recovered, the thermodynamic rarity associated with the materials would decrease, and in turn, the product's sustainability would increase. It is therefore essential to promote ambitious policies for the recovery of valuable materials.

An example of European regulations dealing with the management of waste electrical and electronic equipment (WEEE) is Directive 2012/19/EU of the European Parliament and the European Council, of 4 July 2012. This section includes small and large household appliances, computers, printers, telephones, GPS, photovoltaic panels, fluorescent lamps, etc. This directive seeks to contribute to sustainable production and consumption by reducing the generation of WEEE waste, promoting waste sorting, the correct treatment of waste, as well as waste revaluation and reuse. To this end, a series of minimum objectives have been established and, specifically, as of August 2018, it has been decided that 85% of large devices with an external dimension greater than 50 cm must be evaluated (household appliances, computer equipment, lighting fixtures, musical equipment, etc.) and 80% must be reused and recycled. However, these recycling rates focus only on the recovery of the simplest elements to extract, such as ferrous and some non-ferrous metals (for example, aluminium or copper), or those that are toxic, such as mercury in fluorescent lamps. This results in many other, much more critical, materials being lost along the way.

Nonetheless, some initiatives do exist that seek to recover these metals. In Kosaka, Japan, Dowa Holdings owns a processing plant that allows them to recycle electronic components in computers and telephones to recover base metals such as copper, lead, zinc, as well as other valuable elements such as gold, antimony or indium (Dowa Holdings Co., 2019). They are also investigating in other processes that allow them to recover rare earths, such as neodymium or dysprosium. Other Japanese companies, such as Hitachi, are also conducting experiments to recover rare earths from permanent magnets used in hard drives and air conditioning compressors (Baba et al., 2013). Even so, many materials still end up in landfills or whose functionality is lost after recycling; this situation is much clearer in the automotive sector, a case that we will analyse below.

7.2   Thermodynamic Rarity of Vehicles

Contemporary vehicles have very little in common with those found at the beginning of the twentieth century, and in terms of materials use—the same situation likely applies to future vehicles. At the end of the nineteenth century, the first cars began to be sold as we understand them today. Slowly but surely, the sector evolved from a technological point of view, the value of vehicles increased, and their use spread among the population, helped by the serial manufacturing process. Since the 1940s, engines have been gaining efficiency, vehicles are becoming safer, and performance levels have only increased. In the 1970s, safety and comfort features, such as airbags, seat belts and air conditioning, were incorporated, and systems such as ABS or ESP were included in the 1990s. More recently, electronic systems have also served to offer security, lighting and leisure improvements.

All these changes have generated considerable modifications in the vehicles and in the composition of the different parts used in them. For example, in the case of electronic components, the range of different metals being used is steadily increasing (silver, tantalum, niobium, gallium, indium, etc.). Furthermore, some types of steel used in the suspension arms, wheels, or engine head are being replaced by much lighter aluminium alloys. Rare earth elements such as neodymium and dysprosium have also been incorporated into hybrid cars along with many other metals, such as lithium or cobalt, for electric vehicle batteries.

This has considerably increased the number and diversity of materials contained in automobiles, as we have seen in the power generation sector. From the point of view of the thermodynamic rarity of the vehicle, this growing trend is much clearer (Fig. 7.2), since the materials used today are diverse and, in many cases, critical and difficult to obtain in mines.

If we also take into account that the vehicles of the future are going to be predominantly electric, with the resulting increase in the use of critical materials already been discussed in Sect. 6.4, and heading towards full automation, this growth trend will no doubt continue.

Besides, it should be noted that the vehicle's evolution would not have occurred without the introduction of new and increasingly complex materials, such as ultra-resistant types of steel, which contain elements such as niobium, scandium or molybdenum. These elements have paved the way for the production of vehicles that are faster, stronger, safer, more powerful and are even responsible for lower emissions.

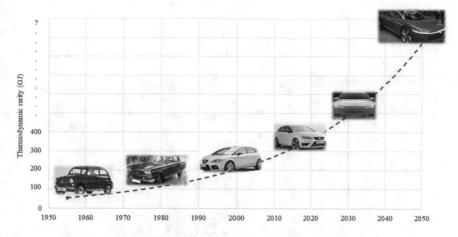

Fig. 7.2   Past and future evolution of the thermodynamic rarity of the vehicle

This new generation of cars will need a more significant number of electronic components, in which metals such as neodymium, praseodymium and dysprosium are used for permanent magnets and other elements such as silver, indium, tantalum or lanthanum (Ortego et al., 2018a). In particular, they will require elements that are fundamental in batteries: lithium, cobalt, nickel, manganese, etc.

That said, the largest portion of a car is composed of steel, aluminium, copper, and plastics, and only a small (yet crucial) fraction of the total is made up of minor but critical and strategic elements. Considering that vehicle sales have doubled in the last 30 years—in the EU, car production has grown by 5% in the last 5 years—this fraction becomes extremely important and strategic for this and other sectors such as renewables. It can therefore be said that vehicles have become de facto "mines on wheels" (Iglesias-Émbil et al., 2020).

Through thermodynamic rarity, we can more accurately analyse a vehicle's physical criticality and assess how the use of these strategic materials is evolving. Initially, we will analyse the thermodynamic rarity of different types of vehicles, after which we will study the various parts that comprise it.

For our first analysis—the difference between mass and thermodynamic rarity of a vehicle—four different models have been taken into account (Iglesias-Émbil et al., 2020). One is an internal combustion engine vehicle (ICEV), running on petrol, and the remaining three are battery electric vehicles (BEVs). Within the latter group, three different types of vehicles have been analysed according to the type of lithium battery they use, all called NCM, made up of lithium together with different proportions of nickel, cobalt and manganese (3:3:3; 6:2:2 and 8:1:1). Figure 7.3 shows the weight of each of these vehicles on the left (in kg) and their corresponding thermodynamic rarity on the right (in GJ).

By weight, iron is logically the most relevant element, being present in many vehicle components, followed by aluminium and copper. We can also observe that

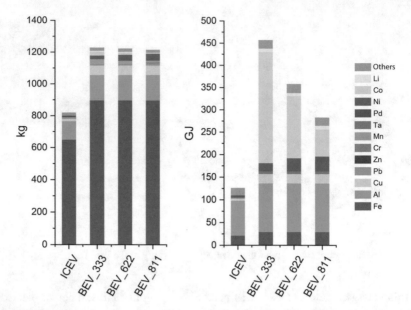

Fig. 7.3 Differences between weight (in kg) and thermodynamic rarity (in GJ) of different models of conventional and electric vehicles. ICEV: internal combustion engine vehicles; BEV_333: vehicle with a 3:3:3 NCM-type lithium battery electric vehicle; BEV_622: 6:2:2 NCM-type lithium battery electric vehicle; BEV_811: 8:1:1 NCM-type lithium battery electric vehicle (Iglesias-Émbil et al., 2020)

the weight of the internal combustion vehicle is somewhat less than that of battery vehicles, which have a similar weight across the different models.

In terms of thermodynamic rarity, ICEV clearly has a much lower value than electric vehicles, mainly due to the components required for the latter's batteries (cobalt, lithium and nickel). Other elements in thermodynamic rarity also stand out, such as tantalum, whose weight was so low that the scale did not allow it to be represented in the figure. With respect to battery electric vehicles, we can clearly observe that the lower the proportion of cobalt, the lower its corresponding global thermodynamic rarity. The amount used of this strategic element is trending downwards.

Having seen the overview of different types of vehicles, let us now look at which components contribute the most to the thermodynamic rarity of vehicles. This will allow us to focus our attention on their eco-design, as we will see in the next chapter.

Specifically, we have analysed the composition of a Seat León III, a vehicle that belongs to segment C and was produced from 2012 to the present. This vehicle weighs 1,270 kg and only 730 kg correspond to metals, the rest belonging to plastics, rubber, glass and textile components (Fig. 7.4). Of the 1,051 parts of the Seat León III, only 794 use some type of metal, which we will analyse below (Ortego et al., 2018b).

To clearly understand the impact associated with the different critical metals, the amount of aluminium and iron present in each car part has not been taken into account. In the analysed model, iron represents 84.3% and aluminium 8.1% by weight of the

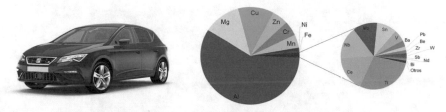

Fig. 7.4 Metal composition in the percentage of the Seat León III (730 kg). Figure developed from data from Ortego et al. (2018b)

metal components of the vehicle. They have been removed from the calculations since these metals are already being recovered in recycling plants (although not quite adequately, as we will see) and, in addition to not being considered critical, their contribution by weight masks the rest of the metals.

To evaluate the criticality for the thermodynamically rare metal content of the selected pieces, the values of thermodynamic rarity (in kJ) and rarity intensity (kJ/kg) are used for each metal and car part. In the first case, the thermodynamic rarity is an intrinsic property of each element. The rarity intensity is obtained by dividing the resulting rarity by the weight of each car part. This allows us to identify those parts that, despite having a very low weight, have a high concentration of critical metals with respect to their total weight.

Once all the calculations for the different parts have been made, they can be grouped into ten different categories based on both indicators: one category for those car parts whose rarity is greater than 1 GJ and nine categories, numbered from A to I, depending on the thermodynamic rarity of the parts and the intensity of the rarity (Ortego et al., 2018c). These divisions have been created by separating the values obtained into five rows and five columns (Fig. 7.5).

The reason why a rarity category greater than 1 GJ has had to be created separately from the others is that some vehicle components, such as the engine, gearbox and front axle, are treated in an aggregated manner rather than separated into small components, so they have a very high rarity compared with the rest and thus must be studied independently.

Of the 794 pieces studied, a total of 31 have been classified as critical in this analysis; the results are summarised in Table 7.3 by criticality categories.

The 31 parts of the vehicle identified as critical have a total rarity of 39.6 GJ, which represents almost 85% of the total rarity of the vehicle, that is, less than 20% of the vehicle components have more than 80% of their rarity.

Table 7.4 shows the list of these 31 car parts identified as thermodynamically critical, along with their rarity and intensity values, as well as the group to which they belong. As we can see, the battery is not included in this list. According to the Directive 2006/66/EC, it must be properly recycled and there are specific treatment centers for this component. For this reason, no additional ecodesign measures are going to be implemented.

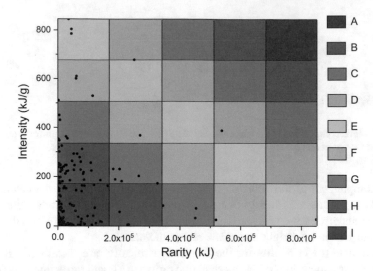

Fig. 7.5 Classification of vehicle components according to their rarity (kJ) and intensity (kJ/g) (Ortego et al., 2018b)

Table 7.3 Number of vehicle parts that belong to each of the categories (Ortego et al., 2018b)

	Number of components
Category > 1GJ	2
Category A	0
Category B	0
Category C	0
Category D	2
Category E	4
Category F	7
Category G	16
Category H	49
Category I	714
Critical components	**31**
Total components	**794**

Now that we have identified the most critical parts of the vehicle in terms of composition, it is time to analyse which parts are actually recycled and which end up in landfills or downcycled in steel or aluminium alloys. That is, they are used to obtain a final product with a lower value than the initial product.

Table 7.4   List of 31 components identified as critical in the vehicle (Ortego et al., 2018b)

	Rarity (kJ)	Intensity (kJ/kg)	Category
Engine	32,099,655	191.52	>1GJ
Gearbox	1,734,110	33.14	>1GJ
Infotainment unit	538,665	389.19	D
On-board supply control unit	250,164	677.90	D
Front axle	849,407	26.81	E
Exhaust gas temperature sensor	44,319	785.39	E
Aerial amplifier (left)	35,217	845.45	E
Aerial amplifier (right)	35,217	845.45	E
Battery	518,856	25.19	F
Combi instrument	269,510	367.46	F
Airbag control unit	113,557	529.44	F
Door control unit (driver)	60,603	610.61	F
Door control unit (passenger)	59,165	600.97	F
Lamp for ambience lighting (driver)	2,148	511.30	F
Lamp for ambience lighting (passenger)	2,148	511.30	F
Generator	453,144	71.54	G
Intermediate exhaust pipe with rear silencer	450,584	32.45	G
Starter	344,093	82.64	G
Wiring	326,124	172.64	G
Rear lighting wiring	265,318	202.30	G
Exterior rear mirror (left)	209,219	180.97	G
Exterior rear mirror (right.)	202,880	230.21	G
Front lighting engine wiring	188,196	188.48	G
Rear screen cleaner motor	182,001	230.49	G
Additional brake light	31,345	354.58	G
Lighting switcher	26,871	362.84	G
Rain sensor	7,124	431.08	G
Air quality sensor	4,922	346.80	G
Speed sensor (front left tire)	4,548	450.91	G
Speed sensor (front right tire)	4,548	450.91	G
Cable shoe used for anti-twist device	1,727	352.43	G

7.3   Loss of Mineral Wealth Associated with Vehicles

If, at the end of their useful life, vehicles were treated appropriately, allowing all their materials to be reused or recycled, the rarity we have calculated in the previous section would not be lost. The automotive sector, in the EU, has been subject to a directive

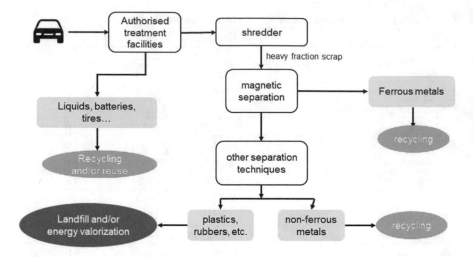

Fig. 7.6 Summary of the stages a vehicle goes through at the end of its useful life (Ortego et al., 2018a)

since 2000 regulating the management of end-of-life vehicles, which establishes that, as of 1 January 2015, the reuse and recovery of an end-of-life vehicle must be at least 95% by weight, and that reuse and recycling must be at least 85% by weight. To meet these figures, vehicles must undergo an optimal treatment and recycling process. It seems that we can relax, then, since the moment we get tired of our old vehicle, its constituent elements will become part of new vehicles, closing the cycle neatly. Or is that not the case? Perhaps all that glitters is not gold after all.

The process that a vehicle follows at the end of its useful life, at least in Europe, consists of four phases in which several agents are involved—a summary of the process can be seen in Fig. 7.6. In the first of these, the user delivers a vehicle that is sent to an authorised treatment centre, where administrative deregistration is processed. Then, within the centres, all dangerous liquids and elements are removed and stored separately, such as hydraulic oils, fuel, antifreeze, batteries, brake fluid, etc. These residues receive separate treatment by other authorised treatment centres. At the same time, the disassembly and storage of parts that can be reused, such as tires or glass, is carried out. Then the fragmentation stage takes place; in corresponding facilities, the vehicles are crushed to achieve fragments between 20 and 40 cm in size. In the post-fragmentation stage and by using different separation techniques (magnetic, gravimetric, optical systems, etc.), ferrous metals, non-ferrous metals, plastics, rubbers, etc. are recovered. In the case of metallic fragments, these are sent to smelting plants (iron, aluminium, copper, etc.), while non-metallic fragments are energetically recovered or recycled. In this way, after the vehicle is deposited in the authorised treatment centre, theoretically almost 5% of the weight is reused, a little more than 78% is recycled and 3–4% is used for energy valorisation, the rest being deposited in landfills (Observatorio Industrial del Metal, 2010). Of this generated waste, the majority is so-called automotive shredder residue (ASR), a mixture of plastics, rubber and foam.

These very promising figures hide some facts that are imperceptible if the evaluation is carried out in mass terms. Indeed, by weight, over 85% of the vehicle is recovered or recycled, but of the more than 50 different elements that compose a vehicle, only steel, aluminium, copper, lead and the catalyst metals are recycled. What happens then to lithium, cobalt, manganese, nickel, rare earth, gallium, indium, niobium, tantalum, etc., many of which are considered critical to the European Union and other economies? The answer is that they either end up in landfills along with the ASR or mixed in very small quantities in different alloys. Despite the fact that there are numerous regulations on vehicle recycling, which promote the reduction of the environmental impact generated by their waste and the recycling of their components, today these processes are rather focused on isolating dangerous components, selling individual parts such as wheels, batteries or catalytic converters and recycling only very abundant metals such as aluminium and iron (Andersson et al., 2017). The minor elements whose demand is expected to increase are also those with lower recycling rates globally. For example, less than 3% of the lithium ion batteries is recycled (Vikström et al., 2013).

Downcycling can be defined as the recycling process in which the final product has a lower value than the initial product. For example, tantalum can end up in small amounts of steel or aluminium alloys. This valuable element does not confer any special properties to the alloy; in fact, it can even ruin it. So, tantalum, although it has not ended up in a landfill, will have lost its original functionality as a semiconductor material used mainly in capacitors. In a recent study, it was detected that out of 17 metals analysed, only platinum from catalytic converters was recycled in a functional way. That is, it was recovered and used again for the same application (Andersson et al., 2017).

Downcycling in vehicles, and especially in the case of special types of steel, supposes a loss of critical elements that are part of the alloys, since at best their concentration is diluted when melted with other types of steel from other sources or it even gives rise to types of steel with a maximum content of certain elements that are too high, preventing their use in certain applications (Amini et al., 2007). A specific case is that of nickel: approximately 40% of the nickel contained in automobiles is reused in steel sheet rolls due to its nickel content, 40% is downcycled together with other metals and the remaining 20% ends up in landfills (Nickel Institute, 2016). However, per the second law of thermodynamics, losses are intrinsic to any process and some of them are inevitable.

How then can we more adequately assess the material losses that occur at the end of life of a vehicle? Our proposal is to use thermodynamic rarity, which allows us to have a figure for the material loss that occurs throughout the recycling of a vehicle, considering not only the amount of lost material but also its quality.

Below we will show as an example the analysis of the loss of materials by downcycling the vehicle analysed in the previous section: a Seat León III, to which a standard recycling process is applied such as the one described above (Ortego et al., 2018a). In this case, downcycling is calculated as the additional amount of metal that would need to be added to re-manufacture a new vehicle assuming the raw materials used for its manufacture came from scrap obtained at the end of the different phases of the vehicle recycling processs. According to this definition, it is also assumed that all metals end up as scrap metal and we also consider that all the metals in the vehicle are recycled, in line with the European and national directives.

To carry out this study, first, we need to know the metallic composition of our vehicle, disaggregated to the maximum possible level, excluding all the parts containing plastics, foams, crystals or rubber (Fig. 7.4). Tires, batteries and other components that are separated in authorised treatment centres prior to crushing are also discarded.

For simplicity, the different parts of the vehicle have been grouped into ten subsystems in order to carry out the downcycling calculation: engine, fuel tank, and exhaust system, transmission, front axle, rear axle, wheels and brakes, selector mechanism, bodywork, electrical equipment and accessories. In this way, each part can also be classified according to the metal used in the alloys. Here we are analysing the alloys of steel, aluminium, copper and zinc. This classification is based on the type of scrap metal recovered in crushing plants.

Once the composition of the vehicle is known, the composition of the scrap obtained at the end of the crushing process must also be known. Since this scrap metal is usually of different composition due to the mixing and dilution of materials, it can be deduced that it is necessary to add a certain amount of elements to obtain the required mixtures for the production of some parts of the vehicle that need types of steel with a specific composition. For components that require metals other than those mentioned above, such as alloys or components containing silver, gold, cobalt, lithium, tin or tantalum, the loss is assumed to be 100% since either they end up diluted in the scrap metal after mixing, completely losing their initial functionality or they end up directly in landfills.

Therefore, once the thermodynamic rarity of the vehicle and the alloys necessary for production are known, the downcycling rate can be calculated. These downcycling rates will always have negative values since they represent a loss of material from an initial situation to a final one, where the quality of the materials decreases. A very high negative value of a subsystem will therefore indicate that a large part of the metals it contains are not functionally recycled.

Table 7.5 shows the downcycling of the ten vehicle subsystems analysed, expressed both in mass and in thermodynamic rarity. In total, the downcycling of metals by weight is 32.8 kg, which represents the amount of metals that have not been functionally recycled in our vehicle, equivalent to 4.5% by weight. If we take into account the recycling targets established by the European directives, more than 85% by weight of the vehicle is being recycled, so the standards are being met. However, when analysing these figures considering the quality of the metals as well as their quantity, through thermodynamic rarity, we can observe that the situation is completely different. The loss of materials is 21,647 MJ, or 26.9% by the rarity of the vehicle.

In short, of four vehicles that enter the end of life stage, only three are recovered in terms of rarity (Fig. 7.7). So, if the objectives set by the vehicle recycling directive

Table 7.5 Downcycling of vehicle subsystems by mass and thermodynamic rarity (Ortego et al., 2018a)	Mass lost (g/vehicle)	Thermodynamic rarity lost (MJ/vehicle)
Engine	−6,024	−10,906
Fuel tank and exhaust system	−3,480	−763
Transmission	−697	−168
Front axle	−1,730	−678
Rear axle	−828	−314
Wheels and brakes	−2,907	−1,246
Selector mechanism	−114	−43
Body	−11,603	−2,368
Electrical and electronic equipment	−5,170	−4,461
Accessories	−286	−699
Total	**−32,839**	**−21,647**

Fig. 7.7 Out of every four vehicles entering the end of life stage, we only recover three in terms of thermodynamic rarity

also depended on the quality of the materials that are recycled, that minimum of 85% is not being met.

If instead of analysing the results by groupings in subsystems we analyse by metal, the most downcycled metals in this vehicle are aluminium, palladium, platinum, copper, tantalum and gold. In the specific case of aluminium, on the one hand, it has a thermodynamic rarity value higher than that of other metals; on the other hand, the aluminium used in many alloys is not functionally recycled. That is, it is not used as a source of aluminium. The average aluminium content in the steel of the Seat León III is 0.75%, while the average aluminium content in the scrap metal is only 0.08%; therefore, this means that a substantial amount of aluminium would need to be incorporated into the steel for it to have the same properties. One of the factors that influences this is the use of ultra-high resistance alloys in the selected model; this steel requires a higher content of aluminium than other types of steel and is used in various parts of the vehicle, such as on floors or foot rails, which explains this average composition of steel with 0.75% aluminium.

Besides, in the case of palladium and platinum, which are mainly used in catalytic converters that are removed in the early stages prior to crushing, they also appear in the composition of other parts, such as in particle filters or in the engine of the rear screen cleaners, which end up mixed with the rest of the metals after grinding without being selectively recovered.

In summary, the recycling process in the automotive sector, and in general of any product, is designed to recover the largest possible percentage of the object by weight. Still, it neglects minor metals whose relative mass is of little relevance and which nevertheless are extremely critical elements for the energy transition. This analysis, combined with future metal demand using sales projections, shows that certain elements such as tellurium, terbium, nickel, indium, lithium or dysprosium, which are currently not recycled, could become strategic for the sector (Ortego et al., 2020). As we have seen in the previous chapter, avoiding dependence on fossil fuels will imply accepting dependence on these metals, so if new technologies really want to be sustainable, efforts should be focused on dematerialisation, the replacement of scarce elements by more abundant ones, increasing recycling rates, implementing eco-design practices that allow better recyclability of the parts used, etc., In short, solutions allow us to reduce the consumption of mineral resources. In the next chapter, we will look at some of these solutions in detail.

References

Amini, S. H., Remmerswaal, J. A. M., Castro, M. B., & Reuter, M. A. (2007). Quantifying the quality loss and resource efficiency of recycling by means of exergy analysis. *Journal of Cleaner Production, 15*, 907–913. https://doi.org/10.1016/J.JCLEPRO.2006.01.010.

Andersson, M., Söderman, M., & Sandén, B. A. (2017). Are scarce metals in cars functionally recycled? *Waste Management, 60*, 407–416. https://doi.org/10.1016/j.wasman.2016.06.031.

Baba, K., Hiroshige, Y., & Nemoto, T. (2013). Rare-earth magnet recycling. *Hitachi Review, 62*, 452–455.

Dowa Holdings Co. (2019). *Annual report 2019.* Available at: https://ir.dowa.co.jp/en/ir/library/ann ual/main/03/teaserItems2/0/linkList/0/link/ar_2019_e_all.pdf.

Iglesias-Émbil, M., Valero, A., Ortego, A., Villacampa, M., Vilaró, J., & Villalba, G. (2020). Raw material use in a battery electric car—A thermodynamic rarity assessment. *Resources, Conservation and Recycling, 158*, 104820. https://doi.org/10.1016/J.RESCONREC.2020.104820.

Ku, A. Y., & Hung, S. (2014). Manage raw material supply risks. *Chemical Engineering Progress, 110.*

Lui, S. (2019). *Statista.* https://www.statista.com/.

Merchant, B. (2017). *The one device: The secret history of the iPhone.* Little Brown, 416 pp.

Nickel Institute. (2016). *Recycling of nickel-containing materials in automobiles.*

Nijman, S. (2019). *UN report: Time to seize opportunity, tackle challenge of e-waste.* https://www. unenvironment.org/.

Observatorio Industrial del Metal. (2010). El sector de reciclaje de metales en España.

Ortego, A., Valero, A., Valero, A., & Iglesias-Embil, M. (2018a). Downcycling in automobile recycling process: A thermodynamic assessment. *Resources, Conservation & Recycling, 136*, 24–32.

Ortego, A., Valero, A., Valero, A., & Iglesias, M. (2018b). Toward material efficient vehicles: Ecode-sign recommendations based on metal sustainability assessments. *SAE International Journal of Materials and Manufacturing, 11*, 213–227. https://doi.org/10.4271/05-11-03-0021.

Ortego, A., Valero, A., Valero, A., Restrepo, E. (2018c). Vehicles and critical raw materials. A sustainability assessment using thermodynamic rarity. *Journal of Industrial Ecology.* https://doi.org/10.1111/jiec.12737.

Ortego, A., Calvo, G., Valero, A., Iglesias-Émbil, M., Valero, A., Villacampa, M. (2020). Assessment of strategic raw materials in the automobile sector. *Resources, Conservation and Recycling, 161,* 104968.

Stanek W., Valero A., Valero A., Uche J., Calvo G. (2017) Thermodynamic Methods to Evaluate Resources. In: Stanek W. (eds) Thermodynamics for Sustainable Management of Natural Resources. Green Energy and Technology. Springer, Cham. https://doi.org/10.1007/978-3-319-48649-9_6.

Valero, A., Valero, A., & Von Gries, N. (2016). Composed thermodynamic rarity of the materials in electric and electronic equipment. In *ECOS 2016—29th International Conference on Efficiency, Cost, Simulation and Environmental Impact of Energy Systems.*

Vikström, H., Davidsson, S., & Höök, M. (2013). Lithium availability and future production outlooks. *Applied Energy, 110*, 252–266. https://doi.org/10.1016/j.apenergy.2013.04.005.

World Economic Forum. (2019). *A new circular vision for electronics. Time for a global reboot.* In support of the United Nations E-waste coalition.

Chapter 8
Looking into the Future

Abstract There are alternatives to compensate for the exponential increase in raw material consumption. Substitution is the most straightforward one, replacing an element with another that is less critical or more abundant. This situation will be explored for electric vehicles, different renewable energies, printed circuit boards and lighting, as each technology requires a good number of scarce elements. Then, circular economy will also be discussed due to the opportunities it may rise in the energy transition to close material cycles. Recycling could be an essential ally. Still, current recycling rates are far from ideal. We will then explore how much recycling efforts should increase to avoid potential material bottlenecks in the future. Using the vehicle as an example, we will describe what spiral economy is, as according to the laws of thermodynamics, circles can never be closed. The possibility of recovering valuable metals from vehicles and eco-design measures for the automotive sector will be analysed. Finally, urban mining, the recovery of metals from waste electrical and electronic equipment, and asteroid mining will be discussed.

The consumption of raw materials has increased exponentially in recent decades, as their use has surged in ever more diverse applications. On many occasions, we see mixtures of multiple materials or the use of certain metals in such small quantities that it is practically impossible to recover them at a cost that is currently profitable. However, the fact that recovery is not profitable does not mean that it should not be happening or that we should refrain from considering alternative designs that facilitate the easy recovery of parts whose composition may be more critical.

This chapter will detail some of the alternatives currently existing to reduce the pressure on the present and future supply of raw materials, especially some considered critical or strategic for the different sectors that we have seen throughout the book. First, we will discuss substitution, which is limited by the properties and requirements that the replaced elements must meet. We will also discuss the so-called "circular economy", and in particular, recycling, the main way of recovering many metals, even though it entails the inherent impossibility of completely closing the cycle. Finally, we will talk about eco-design measures for products and other possible sources of raw materials beyond conventional ones.

© The Author(s), under exclusive license to Springer Nature Switzerland AG 2021　　　207
A. Valero et al., *The Material Limits of Energy Transition: Thanatia*,
https://doi.org/10.1007/978-3-030-78533-8_8

8.1   Substitution of Elements

One of the alternatives currently being looked at with great interest to reduce pressure on the mining sector and on industry (in relation to the supply of raw materials) is the substitution of some critical elements for others that are less so. Essentially, the intention is to create a strategy that allows companies, economic sectors or countries to not depend so much on critical raw materials that come from third parties. This is, in fact, one of the priority areas of the European Union, which includes looking for substitutes for certain materials used in renewable technologies or electronics, among others (European Commission, 2013). Thus, some EU countries, as well as other regions of the planet, have designed strategies focused on the security of the supply of raw materials.

At a European level, various projects and initiatives have been carried out to reduce dependence on raw materials through substitution. Specifically, the CRM InnoNet project (Critical Raw Materials Innovation Network[1]) proposed the five most promising roadmaps that the EU could follow to reduce or eliminate dependence on critical raw materials in the coming years. However, even though substitution was found to be a viable alternative in reducing demand for some elements, the implementation of such strategies could only be carried out if access to said critical raw materials was not guaranteed.

Another initiative, ERECON (European Rare Earths Competency Network), focused on rare earths, analysing the opportunities to extract these elements in Europe, the efficiency in the use of resources, recycling and the challenges that could exist in supply chains (ERECON, 2015). Through substitution, they discovered that the pressure on extraction could be reduced. Nonetheless, this would not be the only solution necessary to solve the shortage of rare earths, as demand would continue to increase. In addition to substitution, they concluded that actions such as recycling and primary extraction of rare earths on European territory, with great potential in certain regions such as northern Europe, should also be prioritised.

The main materials this type of study focuses on are those that are mainly used in renewables and in electrical and electronic components, such as the rare earth already mentioned, gallium, germanium, lithium, tellurium, etc. Other more general reports also exist which focus on analysing the possibilities of substitution of those materials deemed critical for the European Union (CRM InnoNet., 2013; European Commission, 2017).

In recent years, almost 20 projects funded by the EU have focused on finding substitutes for rare earths in permanent magnets, on reducing the use of certain metals in special alloys, on the development of catalysts that do not require elements from the platinum group, etc.

Speaking of substitution generically carries many issues, since it can be influenced by a number of variables. Not all elements can be replaced by others, or in all applications, without losing their characteristic properties. Therefore, for the substitution to be viable, several requirements must be met: (1) performance must be maintained,

---

[1] www.criticalrawmaterials.eu.

(2) the cost of the substitute material must remain competitive, (3) it must be able to be carried out on a large scale and (4) this substitution should not consist of replacing a critical material with a different critical material (Pavel et al., 2016).

Let us now look at some specific technologies, emphasising the critical materials they use and what the possible substitutes might be. We will analyse the elements of electric vehicles, renewable energies, printed circuit boards and lighting.

8.1.1 Electric Vehicles

Of the many elements that are necessary to manufacture electric vehicles, lithium, nickel and cobalt are among the most critical, along with other elements such as gallium, indium or some rare earths. Specifically, these three elements (Li, Ni and Co) are mainly used in batteries. With the increasing demand for sustainable transport options, the number of batteries that will be needed will increase exponentially, as will the demand for these elements.

Within the batteries, there is a wide range available and that could be used as substitutes for cobalt. The best-known type is lead-acid batteries, but there are others such as lithium and iron phosphate, zinc and lithium oxide and nickel batteries. However, these batteries have lower performance and would need to be so large that it would not be feasible to incorporate them into vehicles.

In the case of lithium, some elements such as calcium, magnesium or zinc can replace it as an anode in batteries, although again, the performance would be affected. In a study analysing various types of electric vehicle batteries, it was found that although they were all of the similar weight, their thermodynamic rarity could vary considerably. In the case of the generations of batteries analysed, the thermodynamic rarity decreased progressively due to changes in composition, with a lower amount of cobalt per battery, as we saw in Sect. 7.2 (Iglesias-Émbil et al., 2020).

Due to the problem of the availability of materials to manufacture all the batteries needed in the future, other initiatives and areas of research are beginning to appear that use other solutions, such as carbon dioxide batteries, graphene batteries, sodium batteries (Lee et al., 2017) or solid-state batteries. However, the latter would also require lithium for their operation (Mizuno et al., 2014).

8.1.2 Renewable Energies

Most of the critical elements used by renewable energies belong to the group of rare earths, as is the case of dysprosium and neodymium, in permanent magnets. The use of germanium, gallium and indium is also important in solar panels. Rare earth permanent magnets are much more powerful, so their size can be smaller, and have replaced other types of magnets in many applications. We could use neodymium or praseodymium, substituting one for the other, or seek ways to optimise the magnets so that they require less raw materials. However, the ideal solution would be a change of technology, using ferrite magnets (iron oxide combined with barium, strontium or cobalt oxide), already used in some electric motors. However, as with many other element substitutions, this would result in a loss of performance. As for germanium, there are several alternatives for some applications, and, in the case of solar panels, the most viable substitute is silicon, although germanium is preferred since it entails a reduction in the size of the electronic parts. In solar panels of satellites, despite research being carried out in this area, there is no material that can replace germanium (European Commission, 2017). A similar situation occurs with gallium and indium, challenging to replace in many applications due to their properties. In terms of indium, there is no viable substitute for the semiconductor compounds used in thin-film solar cells (European Commission, 2017).

Another factor to consider in the case of renewable energy is the fact that, since its production is linked to the existence of wind or sun, a solution would need to incorporate some energy storage system. This would allow the generated energy in any given time to be used immediately and at a future point in time while helping to maintain a stable electrical network in the event of overproduction. This storage can be carried out using batteries that, in turn, will contain critical raw materials. One of the largest projects for energy storage from renewable sources using batteries is in Australia, on the Hornsdale wind farm, with a large 100 MW (129 MWh) lithium-ion battery that has been operational since 2018.[2]

8.1.3   Printed Circuit Boards

Printed circuit boards are used to connect different electronic components that are usually composed mainly of copper, fibreglass, ceramics, resins, etc. (Fig. 8.1). However, in addition to these materials, there are many others in small quantities that are scarcer, among which we find antimony, beryllium, gallium, germanium, indium, niobium or tantalum.

Antimony is used as a dopant for n-type silicon semiconductors in small amounts, and other dopants that can carry out the same function are phosphorus and arsenic. The copper and beryllium alloys used are six times stronger than if only copper was used. It is also used as an insulator between the silicon chip and the metal parts in high-power devices (CRM InnoNet., 2013). In some applications, beryllium can be replaced by magnesium, nickel, silicon or titanium, albeit with a loss in performance. Gallium and germanium are used in transistors; gallium can be partly replaced by aluminium or indium, and germanium by silicon, but the properties of the resulting products are not the same. Finally, tantalum is mainly used in capacitors (Fig. 8.2); before using this element, aluminium was used, and today there are some ceramics that can fulfil similar functions. In fact, studies have been carried out on what it would cost to recover these scarcer elements in printed circuit boards, concluding that the

---

[2]https://hornsdalepowerreserve.com.au.

Fig. 8.1 Motherboard of a laptop (Lifetec LT9303). Author: Raimond Spekking. CC BY-SA 4.0. Wikimedia Commons

Fig. 8.2 Tantalum capacitor on a printed circuit board. Author: Epop. Wikimedia Commons

cost would be less than the market cost of the metals present in them (Talens Peiró et al., 2020).

8.1.4   Lighting

In lighting, rare earths are used in fluorescent lamps; for example, terbium emits green light, and europium can emit red or blue light. In these cases, there are no viable substitutes, although they are losing importance compared to LED lighting. Instead, this LED technology uses indium, gallium and germanium, elements for which there are no substitutes in these applications, although options such as using zinc oxide or magnesium sulphide to replace gallium are being considered. However, there is a different technology that does not require these elements for lighting, OLED technology. In this case, those responsible for emitting light are organic diodes that react to various electrical stimuli, generating and emitting light independently. On the negative side, these technologies have a shorter lifespan, are somewhat more expensive to manufacture today, and their recycling process is not straightforward (CRM InnoNet., 2013).

Table 8.1 summarises the technologies along with some of the elements used in them and the possible substitutes that have been discussed. Even though many of these elements are critical, there are few viable alternatives that exist today to replace them, although some efforts are being made in the specific case of batteries for electric vehicles.

There are various types of substitution, categorised according to their purpose. First, we have pure substitution, which simply implies exchanging one substance for another for a specific application. Another type is to switch a service for a product, more focused on extending the useful life of a product and its intensity of use; for example, through rental or leasing networks. Third, the substitution of one process for another, reflected in a change in how a product is manufactured. Finally, we can switch substances for other, more advanced technologies (CRM InnoNet, 2015).

In the cases analysed, the alternatives focused almost all on the first category of substitutions: the pure substitution of some elements for others. Only in very few

Table 8.1 Summary of some of the elements used in different technologies and possible substitutes

Technology	Elements used	Possible substitutes
Electric vehicles	Li, Co	Ni, Ca, Mg, Zn, graphene
Renewable energy	Nd, Pr, Ga, Ge, In	Ferrite, Si
Printed circuit boards	Sb, Be, Ga, Ge, In, Nb, Ta	P, As, Mg, Al, Ni, Si, Ti, In, ceramics
Lighting	Eu, Tb, In, Ga, Ge	Zn, Mg, organic compounds

cases do we see substitutions of substances for other technologies to reduce the use of critical materials, such as replacing LED technology with OLED.

Today, and with existing technology, critical raw materials are very difficult to substitute in various applications without loss of properties or performance, and most of the possibilities being analysed are still in the early stages of development. Thus, the real-world application could still be 5–15 years in the future, according to some studies (Bouyer, 2019).

There is still much research work to be done to decrease the number of critical materials used in products and processes. It is also important to promote initiatives that promote the best use and exploitation of resources, through, for example, the so-called "circular economy" and in particular recycling, discussed in more detail below.

8.2 The Circular Economy

The concept of circular economy is booming in today's society. It refers to a type of economy that seeks to achieve sustainable and economic environmental growth, maintaining the value of products, materials and resources for as long as possible. In other words, it seeks to reuse goods a maximum number of times within the system, thus minimising losses and reducing pressure on the planet, in particular the mining sector in charge of supplying us with raw materials.

There is no clear definition of the circular economy. Indeed, the circular economy has been influenced by economic theories of over half a century ago, especially those associated with industrial metabolism and the flows of matter and energy (Jiménez Herrero, 2019). The most widely accepted definitions come from the Ellen MacArthur Foundation and the European Commission. The central idea of all of them resides in redefining the traditional economic system based on the linear model of produce-use-throw to a regenerative one where products and services are commercialised in closed cycles.

According to the Ellen MacArthur Foundation, the principles on which the circular economy is based are (EMF, 2015):

1. Preservation and improvement of natural capital, controlling finite stocks and balancing the flow of renewable resources, circulating products, components and materials with maximum utility at all times, both in technical and biological cycles.
2. Optimisation of the use of resources, always circulating products, components and materials at their highest level of utility at all times, both in technical and biological cycles.
3. Promotion of system efficiency, revealing and eliminating negative externalities.

These pillars form the basis of a strictly circular economy that would involve designing without waste; increasing resilience through diversity; working towards the use of energy from renewable sources; thinking about systems and understanding the links between them and their consequences; promoting prices that reflect real costs by introducing negative externalities and eliminating perverse subsidies; and thinking of cascading, diversifying reuse along the value chain, as shown in the now-classic figure of the Ellen MacArthur Foundation (Fig. 8.3).

As stated by Kirchherr et al. (2017), the circular economy is operationally identified with the expanded 3Rs principle (reduce, reuse and recycle), such as the one proposed by the Dutch Environment Agency (PBL) of the 9Rs: reject, rethink, reduce (related to the use and production of smarter products), reuse, repair, renew, remanufacture, re-adapt (in order to extend the useful life of a product and its parts) and, finally, recycle and recover (in the last step in the hierarchy) (Potting et al., 2017). The circular economy could also be called Re-economy (Valero and Valero, 2014). The prefix Re- means to return to the initial condition and repeat or restore an action continuously. Most "re"-verbs can be adopted in the circular economy: re-design, re-invent, re-organise, re-vise, re-refine, re-create, re-consider, re-know and so on. All of them are intimately linked to the ethics of the finiteness of natural resources.

Although located in the last step of the chain, recycling is probably the main activity that we associate with the circular economy and is undoubtedly an important measure to reduce the extraction of raw materials. Let us see below the current status of recycling as well as its limitations.

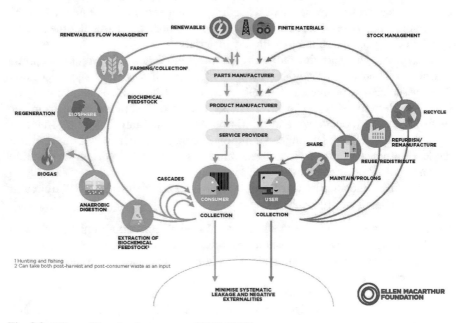

Fig. 8.3   Pillars of the circular economy (EMF, 2015)

8.3 Recycling

In the case of mineral and metal resources, recycling, as already happens with plastics and paper, could be one of the solutions to creating a secondary source, substituting its extraction from mines. The existence of these secondary sources of metals may contribute to the development of the circular economy and reduce the risk of supply from third countries.

Recycling is also a measure that can help reduce the environmental impact associated with these metals, since the energy required to recover a metal may be less than the energy required to extract the same amount of metal from a mine. That said, the contribution of recycling for many metals today is considerably less than would be expected, and only a few are effectively recycled. Indeed, an immensity of products that reach the end of their useful life does not end up in treatment plants or recycling centres. Instead, they end up stored or, worse still, in uncontrolled landfills. There are also a series of materials that take a long time to be reused again due to the long life of the objects that contain them, such as vehicles, wind turbines, buildings, etc. It is therefore important to consider that not everything can be recycled at the same rate or in the same way.

One of the most functionally recycled elements worldwide is aluminium, largely due to the selective collection of metals. The collected aluminium is melted and transformed into secondary aluminium ingots, which are then used for the production of rolled aluminium, construction plates, aluminium foil, etc. The energy saving in this process is extremely high when compared to the bauxite aluminium extraction process. Some sources, such as the Aluminum Association,[3] assure that savings are greater than 92%. However, the final quality of secondary aluminium should be considered for the comparison to be rigorous.

A historical analysis of primary production of aluminium metal and secondary production from recycling showed that, although secondary aluminium production has only increased since the 1950s, it has barely represented 35% of the total available aluminium (Gerber, 2007).

Looking closely at Fig. 8.4, we can see that during the 1950–2020 period, the production or demand for aluminium experienced an annual growth of 2%, while the growth in the demand for recycled metal grew at a rate of only 0.25%. If this trend was to continue in the future, through a simple calculation, it would take over 250 years to reach a recovery efficiency of 98%. Demand, on the other hand, would multiply by two every 40 years and, if there was no recycling, primary extraction would multiply by two every 35 years. Thus, even if we reached this ideal recovery rate of 98%, extraction and demand would continue to increase following these numbers, no matter how much aluminium were recyled. The implications of this situation are clear: the only way to replace primary extraction is to reduce demand below the recycling rate. Mathematically speaking, this would imply that the current growth rate of demand should be below 0.25%. However, as we will see later, recycling rates close to 100% are more an ideal than an achievable reality, even in the case of much

---

[3] www.aluminum.org.

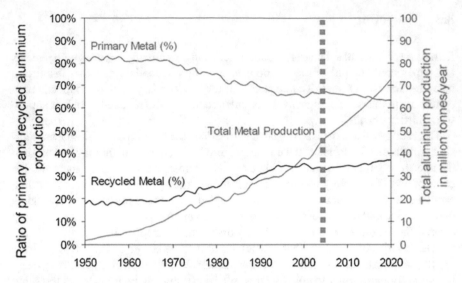

Fig. 8.4 Historical production of primary and secondary aluminium (recycled). Reproduced from Gerber (2007)

more valuable metals than aluminium, where the incentives to recover such metals are higher.

When talking about recycling rates, we usually refer to metals including iron, copper, lead, zinc, nickel, titanium. One of the fundamental questions we must ask ourselves is where these materials come from. Broadly speaking, iron and other ferrous metals come from the recycling of types of steel and iron. Non-ferrous metals, such as aluminium or copper, come from other sources and, although they are lighter in mass, they generally have a higher economic value.

In both types of metals, ferrous and non-ferrous, scrap metal can come from three different sources. First, there is the scrap metal originating from factories, refineries or smelters, which is usually recovered and reused in the same plant. Second, there is the scrap metal of industrial origin, coming from cuts arising from the manufacturing of a product, such as household appliances, vehicles, machines, etc. Finally, there is scrap metal from goods that have already reached the end of their useful life. For example, a vehicle has been taken to a junkyard or treatment plant. This represents, in percentage, the largest quantity out of the three sources. It is precisely in this last part where the different policies and laws on the proper treatment of waste are focused, forcing the collection and recovery of a large part of these devices and products.

On many occasions, the "recycling" policies remain in the field of mere collection. For example, it is common to speak of a certain recycling rate for glass, plastics or cardboard, when in reality, the figures refer to the selective collection of this waste. The actual percentage that is recovered and reintroduced into the system is probably much lower than advertised figures. The causes of this are multiple, including lack of cheap and efficient technology to properly separate raw materials, non-competitive costs compared to the use of virgin materials, reduction of the quality of recovered materials not suitable for use in original or equivalent value (ending up downcycled at best), etc.

Unfortunately, there are very few studies at the global level of the effective recycling rates of metals, since, in many cases, they vary depending on their applications, the area where they have been used, and the time of use. The iron used in vehicles is different to that used in the construction sector, or the aluminium used in a soda can or in a vehicle, since the time that the element remains in use is much higher in one case than in another. However, there are also numerous differences depending on whether we're discussing these two base metals or rarer elements such as rare earths or lithium. Thus, it is easier to find figures for the recycling rates of elements such as iron, aluminium, copper, zinc and tin, which have shorter lifetimes and are easier and more economically recoverable.

Furthermore, when it comes to metal recycling rates, several types can be distinguished depending on the limits of the system that are considered. For example, one of the most common is the end-of-life recycling rate, known as EOL-RR, calculated by the ratio of the amount of scrap recycled to the metal content of the products. Another way of calculating recycling rates is using recycled content, which measures the actual amount of metal entering the industrial life cycle of a particular item.

In relation to this last classification, the United Nations Environment Program (UNEP), produced a report in 2011 on the recycling rates of different elements of the periodic table, with particular emphasis on base metals and critical metals (UNEP, 2011). Although this report is now a few years old with no update published since, many of the metals included in it were studied for the first and only time in detail, so the figures obtained are of great value.

Figure 8.5 shows a summary of the recycling rates (recycled content) of a total of 54 elements in the periodic table; those that appear in grey were either not analysed or insufficient information is available. As we can see, very few elements have a recycling rate of more than 50%: these include lead, iron or titanium. It is striking that the recycling rate of niobium, platinum and palladium is so high: this can be explained by looking at the sectors in which they are used, vehicle alloys and catalysts, respectively, which typically carry out recycling with a certain level of efficiency. At the other extreme, of the items marked in the figure, almost half have a recycling rate of less than 10%, and that value in some cases does not even hit the 1% mark. The clear examples are rare earths, tellurium or lithium, critical elements for renewable technologies, demand for which has sky-rocketed.

A report prepared by the European Commission analysed the recycling rates of materials considered critical, obtaining somewhat different results for some substances (European Commission, 2018). Although many of these critical materials have a high potential for recycling, rates are generally much lower than expected. For example, the contribution of recycling to the demand of the EU in the case of tungsten is 42%, 28% for antimony, and just 2% in the case of germanium.

Comparing the recycling rates of other metals, such as aluminium, at a global level, this figure is around 36%, while in Europe, it is still only 12%. In many of the elements of the periodic table above, recycling rates in Europe are lower than

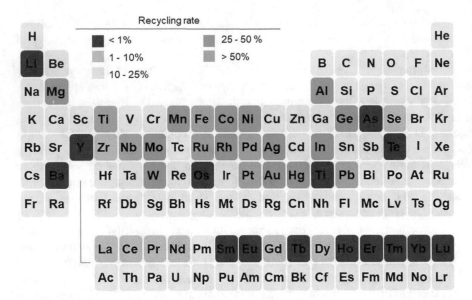

Fig. 8.5 Summary of the recycling rates of some elements in the world. Source: UNEP (2011)

their global counterparts, except for silver, nickel, antimony or tin, whose rates are higher than their global equivalents. Significantly, in Europe, the recycling rates of europium and terbium are extremely high compared with global recycling rates, which are almost negligible (<1%). Specifically, in 2013, a total of 33.5 tons of europium were functionally recycled in Europe and almost 22 tons of terbium, both elements mainly from the recycling of fluorescent lamps (BIO by Deloitte, 2015).

Increasing functional recycling rates would undoubtedly ease tensions in the supply of critical raw materials to develop new technologies. In the study we mentioned in Sect. 6.2, regarding bottlenecks that could appear in the future manufacturing of renewable technologies (Valero et al., 2018), we evaluated how much recycling rates should increase compared with the current rates to avoid supply shortages of certain raw materials (Table 8.2). It was assumed that primary extraction would continue to increase, following the so-called Hubbert curves explained in Sect. 5.2.

The highest annual growth of these recycling rates corresponds to lithium (4.6%), chromium (2.5%), cobalt (1.8%) and cadmium (1.3%). The case of lithium is highly relevant as its demand is expected to grow in the future due to its importance in energy storage and its low recycling rate, which today barely reaches 1%. It is also striking how small the efforts would be to prevent the emergence of certain bottlenecks in manganese, neodymium or tin, with increases in the recycling rate of less than 0.1% per year. Naturally, if these recycling rates were higher, the primary extraction of minerals could be diminished or at least its growth attenuated, as we have seen with aluminium.

Table 8.2 Current recycling rates, annual growth and recycling rates in 2050 to prevent annual demand from exceeding annual production (Valero et al., 2018)

	Current recycling rate (%)	Annual growth (%)	Recycling rate in 2050 (%)
Ag	30	0.6	37
Cd	25	1.3	39
Co	32	1.8	59
Cr	20	2.5	47
Dy	10	0.9	13.7
In	37.5	0.5	44.7
Li	1	4.6	4.8
Mn	37	0.7	38
Mo	33	0.7	42
Nd	5	0.1	5.2
Ni	59	1	41
Se	5	2	10
Sn	22	0.1	22.8
Ta	17.5	0.1	18.2

Unfortunately, of the metals that began to be used industrially only a few decades ago, such as rare earths, tellurium or lithium, among others, there is very little information on the efficiency of recycling processes. Furthermore, some of these elements have properties that make these processes technologically and economically very complex. Besides, recycling processes have their own thermodynamic limits: as it is technically impossible to achieve 100% recycling of any element, there will always be losses in the system, leaving dispersed metals behind, accumulated in landfills or diluted in scrap metal, as we saw in Sect. 7.3 in the case of vehicles.

8.4 The Thermodynamic Impossibility of Closing Cycles: The Spiral Economy

Unlike fossil fuels, minerals, once used, are not lost. In theory, they could be reused repeatedly. However, as we have seen in the previous section, we are extremely far from achieving 100% recyclability for all elements on the periodic table. The highest rate we have reached, that of lead, barely exceeds 60% globally. Some of the causes have already been mentioned, but the most important is related to the technical inability to recover each and every one of the elements used in the technosphere. In this case, the fault lies not only in the lack of adequate technology, which is certainly far from optimal. The causes are deeper and relate to thermodynamics, in particular to irreversibility and the concept of entropy.

An irreversible process is one that could regain the initial state from which it has started, but it will do so at the cost of needing much more energy than that released in the outgoing process. An ice cube melts spontaneously, but if we want to form it again, we will need, in real life, much higher energy than that which was necessary to liquefy it. As with the ice cube, all spontaneous processes are irreversible, and returning to the initial state is very costly. So reversibility is practically unattainable, although it serves as a reference to quantify how far we are from ideal behaviour, which will be measured with the entropy generated from the processes in units of energy divided by temperature (kJ/K) or, more comfortably, with the exergy in kJ. The more irreversible a process is, the more entropy will have been generated, or the more exergy will have been destroyed. So unrecyclability is actually a particular case of irreversibility. The question that arises then is, what is the physical cost of recovering materials? A pure substance with low entropy, when mixed with other substances to create a product, increases entropy, with a loss of quality (exergy). If the process was ideal, that is, reversible, we would be able to fully recover materials with little effort. Naturally, reality is far from that chimera and recovering its state of purity can only be done at the cost of much more exergy (useful energy) than that which was lost in the mixing process. The unrecyclability of materials is therefore related to the generated entropy associated with the manufacturing of that product. Entropy, a measure of disorder, in turn, depends on the number of components of the mixture (the product) and the relative amounts of each element (concentration).

Furthermore, the relationship between separation and concentration costs is not linear. Purifying 99.9% of a substance costs exponentially more energy than purifying up to 99% of that substance. It is therefore very important to stay in the exact specifications of the process/product, because going over implies an increase in costs. Maintaining quality is synonymous with efficiency. It is also vitally important to avoid mixing more than necessary. In this sense, recycling can go in the opposite direction of creating stronger materials with superior properties.

In the journal Nature, the new metallurgical technology of "metallic mixology" is described (Lim, 2016). This article presents examples such as that of Zhang from 2009, which described an alloy composed of cobalt, chromium, iron, nickel and aluminium that was 14 times stronger than pure aluminium, but whose ductility was almost three times greater—a measure of the ability of a metal to work without breaking. For his part, Yeh devised a cobalt, chromium, iron and nickel alloy that can cool below liquid nitrogen temperatures without becoming brittle. Cem Tasan, a metallurgist at MIT, mixed iron, manganese, cobalt and chromium in an alloy that is extremely hard and highly ductile. Other investigations aim at alloying elements with niobium, tantalum and chromium to obtain alloys with very high melting points. All these examples are so-called "high entropy alloys". This bombastic name refers to the downside of these super-alloys: the technical impossibility of recycling them.

A common mistake that occurs in the industry is to mix effluents to dilute the polluting effects of a certain substance. This process, far from remedying the evil, will irreparably increase the entropy of the whole. Polluting flows must be segregated, never mixed.

Unrecyclability is also related to the nature of the materials that have been mixed and this is where metallurgy comes into play. We saw in Sect. 7.3 above that, with the physical separation processes, we can only aspire to recover ferrous metals—mostly aluminium and copper. To go further and recover minor but valuable metals, we must employ metallurgy.

The so-called "wheel of metals" developed by Markus Reuter et al., helps to understand the material losses that occur in any metallurgical process aimed at recovering minor metals from a product. The inner circumference of the wheel indicates each metallurgical route for obtaining the major metals, divided into 10 different metallurgical groups: iron and steel, aluminium, titanium, magnesium, lithium, rare earths, tin, copper and nickel, zinc and lead, manganese, nickel and chrome. Each of these metallurgical routes will be made up of different processes from which we can obtain other minor elements specified in green and located in the concentric circles, in addition to the main metals. The elements shown in yellow are those that end up downcycled in some type of alloy, while the red ones are those that are irretrievably lost.

The first lesson to be learned from the wheel is that recycling rates strongly depend on the composition of the input materials to the metallurgical process. So, in the recycling of a certain product, it must be decided at first which are the metals that are to be recovered to choose the metallurgical route to apply.

Let us imagine that we choose the metallurgical route of steel. We will have to assume then that any copper impurity entering the recovery process will be lost, as can be seen in the wheel of metals (Fig. 8.6). To prevent this from happening, the ideal solution would be to disassemble the product to the maximum to ensure that each fraction goes to the appropriate process and does not lose minor metals (Reuter et al., 2013).

8.4.1   Recovery of Valuable Metals in Vehicles

What if we could properly apply a recycling process to vehicles? Let us suppose that instead of fragmenting them and using the physical processes in use today, we could properly disassemble their most important parts to ensure they pass through the metallurgical process that maximises the recovery of valuable metals. This was one of the objectives of the AWARE project, carried out jointly for SEAT by the CIRCE Institute of the University of Zaragoza and the Helmholtz Institute Freiberg for Resource Technology (HZDR).

With this purpose in mind, 11 parts with a high thermodynamic rarity were selected and the vehicle disassembly times were evaluated, as well as the number of models that shared each part (Table 8.3). Let us remember that thermodynamic rarity assesses the physical criticality of the materials in each object but does not take into account how these metals appear in the vehicle, since metals can appear dispersed in different parts or concentrated in one. The way the metals appear logically affects their recyclability and ease of separation, but whether they are concentrated or dispersed, the total thermodynamic rarity of the vehicle will be the same.

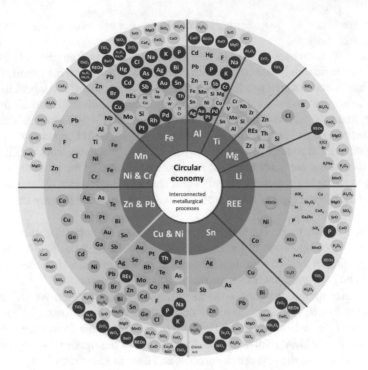

Fig. 8.6 Wheel of metals showing the difference in metallurgical processing of different base metals. The wheel shows on green the elements that are usually recovered, in yellow those that end in slags or alloys with some functionality and in red those that are lost (Verhoef et al., 2004)

Table 8.3 Data summary of the main selected car parts

Piece	Number of units/year	Rarity density (kJ/g)	Disassembly time (s)
Engine	3,557,862	71.54	1,048
Infotainment unit	545,520	386.19	617
Combi Instrument	196,733	367.46	665
Additional brake light	1,375,741	354.48	324
Exterior mirror	393,466	230.21	70
Air quality sensor	3,551,770	346.8	830
Cable shoe used for anti-twist device—ground line	487,833	352.43	1,237
Rain sensor	6,385,923	431.08	1,233
Speed sensor and ABS sensor	14,590,752	450.91	600
Battery wiring	160,900	319.98	Low
Vehicle side	321,800	39	High

Fig. 8.7 General exploded view of the dashboard of a vehicle

These 11 car parts were then sub-disassembled until they could be divided into main fractions: ferrous fraction, non-iron fraction, aluminium and rest. Figure 8.7 shows, for example, the general exploded view of the dashboard.

For the three metal fractions, the metallurgical processes necessary to obtain the maximum of minor metals were designed with the HSC Chemistry software. Figure 8.8 shows a summary of the processes that apply to the refining of copper and its by-products (non-metallic fraction). The metallurgical process is very complex,

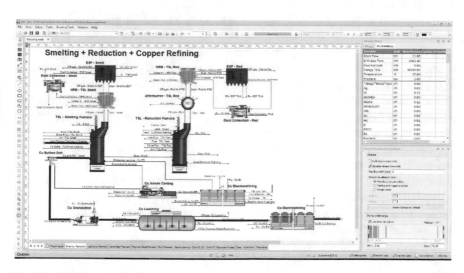

Fig. 8.8 Flowchart for non-ferrous metals

Table 8.4 Summary of the recovery percentage of the non-ferrous fraction

Element		Element		Element		Element		Element	
Aluminium	●	Antimony	●	Arsenic	●	Barium	●	Bismuth	●
Boron	●	Cadmium	●	Carbon	●	Cerium	●	Chromium	●
Cobalt	●	Copper	●	Dysprosium	●	Gallium	●	Germanium	
Gold	●	Hydrogen	●	Indium	●	Iron	●	Lead	●
Lithium	●	Magnesium	●	Manganese	●	Molybdenum	●	Palladium	●
Phosphorous	●	Platinum	●	Ruthenium	●	Silicon	●	Silver	●
Strontium	●	Tantalum	●	Terbium	●	Tin	●	Titanium	●
Tungsten	●	Vanadium	●	Ytterbium	●	Yttrium	●	Zinc	●
Zirconium	●								
● 0-25% Recovery		● 25-50% Recovery		● 50-75% Recovery		● 75-100% Recovery			

including 50 different elements, 11 flow diagrams, 124 unit operations and 524 individual flows.

This very complex process results in a significant recovery of many minor metals, as shown in Table 8.4. That said, not all metals will be fully recovered. We will be able to recover most of the copper, cobalt, gold or platinum. Others, like aluminium, boron, tantalum to name a few, will end up getting lost in the slags of the process.

A similar procedure can be performed for the recovery of steel and aluminium. In these cases, the process is much easier. With respect to steel, the process consists of melting the scrap metal in an electric arc furnace (EAF). Impurities that have oxides more stable than iron are collected in the slag, with elements such as aluminium, magnesium, or calcium being largely removed. Molten steel is further refined through a ladle furnace to remove impurities such as oxygen, sulphur or carbon. However, impurities such as copper or nickel cannot be removed because they are stable in the metallic form under EAF operating conditions. Consequently, high-quality alloys may need to be diluted with high-quality scrap or pure iron to match the specific composition of the alloy.

Aluminium recycling is made up of the melting and refining stages. In melting, aluminium scrap melts together with a slag flux that protects molten aluminium from oxidation. The high reactivity of aluminium makes it very important to control impurities in the feed, as they can react with aluminium and generate material losses. In the refining stage, magnesium impurities in the molten aluminium are decreased.

For both aluminium and steel, it is important to remove as many impurities as possible, otherwise, the recycled material will have to be diluted with more steel or primary aluminium to achieve the required specifications.

Mixing all types of steel and all aluminium in smelting furnaces results in very low-quality steel and aluminium. For example, the side of the vehicle contains 40 different types of steel, some of them of very high quality. In the recycling process, all types of steel are mixed, resulting in low-quality steel. To meet the requirements of the steels included on the side, we need to incorporate a huge amount of virgin raw material into the mixture.

In short, even by disassembling the most critical parts of the vehicle as much as possible by applying the best possible recycling metallurgical technologies to these, the complete and high-quality recovery of all its constituents is physically impossible.

The more complex a part (of a car or any other object) is in terms of the quantity of materials and mixture, the more difficult its total recovery will be. From a recycling point of view, these types of products are commonly called "monstrous hybrids", a term coined by Michael Braungart and William McDonough (McDonough and Braungart, 2002). In this group, we find of course the high entropy alloys that we have mentioned, but also the humble Tetra Brik milk carton, composed of plastics, cardboard and metals that are impossible to recover in practical terms.

In short, the circular economy, an objective to which we must aspire, is nevertheless an impossible chimera to achieve, simply because of the limitations imposed by the second principle of thermodynamics and the remoteness of our technology to effectively and efficiently recover the different materials that make up the products currently used. A more accurate term would be "spiral economy" instead if we accept that thermodynamically speaking, circles can never be closed.

8.5 Eco-design Measures

Another currently viable option to promote the efficient use of resources and optimal recovery is eco-design, that is, the design and development of products so that they are durable and incorporate circularity as opposed to linear thinking. It should be borne in mind that the products are associated with numerous impacts related to their life cycle, from the extraction of raw materials in mines to their elimination and recycling. And the more advanced a product is in its production chain, the greater the loss of resources it will suffer when it is discarded and its materials dispersed in Thanatia (higher exergy cost). Thus, during a product's life cycle, it must be possible to recover the materials it contains or the resources necessary for its manufacturing, such as metals, for later reuse. This not only decreases and controls the environmental impact they generate but can also help prevent primary extraction and depletion of mineral resources.

Therefore, in the design of any product, it is essential not only to think about its production and use phase but also about its end of life. No product should become a waste. In fact, any residue symbolises a failure in system design. On the contrary, nature does not produce waste since the waste of a certain organism constitutes the raw material of others, thus closing the cycles. We must therefore learn from nature and practice what is called "biomimicry" (Benyus, 2002).

This reasoning is in line with the "Blue Economy", which proposes the use of resources through cascading systems, in such a way that one product constitutes the resource of another product and so on (Pauli, 2010). Another classic reference book is McDonough and Braungart's *Cradle to the Cradle* (2002), which also discusses taking a biomimetic and regenerative approach to product and system design.

Like products, industrial systems can be eco-designed to optimise the use of resources. This is the case of eco-industrial parks (Frosch and Gallopoulos, 1989), which apply the principles of industrial symbiosis (Chertow, 2007) based on the original concept of industrial metabolism defined by Ayres (1994). In the industrial symbiosis, the waste from one industry becomes the raw material of an adjacent one. The most prominent case of an eco-industrial park is that of Kalundborg in Denmark.

Returning to products, these must also be designed to extend their useful life as much as possible and to remain in the technosphere for a long time. Durability or "lengthening" as opposed to "planned obsolescence" must be a fundamental requirement of all products. Designing to last means using fewer materials with the highest quality, robustly built and with high reliability, which are easily disassembled, repairable and modular for the eventual replacement of parts (Cramer, 1997). Today, there are no standards that make the life of repairers or recyclers easier. Each model, even from the same brand, has its peculiarities. There are no instruction manuals for anyone who wants to repair or recover parts of an object that no longer works. In addition, the tools necessary to disassemble the equipment vary from one object to another, which makes the world of recycling a very artisanal one.

To this, we must also add that nowadays the production of new objects is usually cheaper than the repairing of old ones. The culture of disposable products has been strongly established in society as raw materials and the labour force relegated mostly to developing countries are still too cheap to compete with a culture of reparation. In this sense, fashion is one of the sectors that consume the most resources. As explained by Allwood et al. (2006), sustainable fashion and its promotion constitute a huge challenge for the future.

Another fundamental aspect in the eco-design of products is dematerialisation. A better product will be one that uses fewer materials, less water and less energy to manufacture. Of course, we will start by removing those most critical materials from the product (in our case, those with the highest thermodynamic rarity). Wherever possible, it will be interesting to miniaturise objects as this will help reduce the packaging and transport costs. However, miniaturisation can lead to a substantial reduction in recyclability, so both factors must be compared when it comes to eco-design. Unfortunately, dematerialisation will always be partial since human beings need materials and energy for their survival. Nonetheless, the amount used can be dramatically reduced even by a factor of 10, as suggested by the "Factor 10 Institute" and the "Rocky Mountain Institute" (Schmidt-Bleek, 1994). Factor 10 is derived from the ideas laid out in the Club of Rome's Factor 4 report (Weizsacker et al., 1998), which proposed how to increase wealth by halving the resources used.

All this being said, certainly, the most important measure to reduce the consumption of resources is through the promotion of a service-based economy, where what is sold is the service offered by a product rather than the product itself. Through "eco-leasing", companies will be able to monitor their products throughout their lives, promoting design for repair, reliability and lengthening.

8.5.1   Eco-design Measures in Relation to Vehicles

Given that the automotive sector is one of the main consumers of raw materials, and this trend is expected to accelerate over time due to the introduction of electric vehicles and growing demand, it is an interesting sector to analyse in-depth.

Automobile manufacturers have focused in recent decades on improving the environmental performance of their vehicles through measures related to decreased emissions and fuel efficiency. As explained in Sect. 7.2, the recycling quotas established by the European Union (which require that 85% of the vehicle be recycled *en masse*), are a starting point, but meeting these requirements does not guarantee a sustainable use of raw materials or of mineral resources. Furthermore, it encourages the recovery of abundant metals, such as iron or aluminium, at recycling points, but leaves aside the recovery of other much more critical metals such as gold or tantalum, which end up being downcycled or dumped in landfills.

We observed in Sect. 7.3 that the current processes of recycling vehicles through fragmentation and separation by physical means implied the loss of vast amounts of valuable metals. Let us recall that, in practice, of four vehicles that are recycled with current processes, one is lost if measured in terms of thermodynamic rarity. On the contrary, if we apply more selective recycling using metallurgical processes, the recovery of minor metals is much higher. That said, a not inconsiderable percentage is also lost due primarily to the way the vehicle parts are designed.

Increasing the recyclability of existing vehicles involves segregating materials to avoid increasing the entropy of the mixture. Currently, recycling is based on a crushing process of the vehicle that has only had the wheels, liquids, battery and perhaps some valuable and easily disassembled parts removed for reuse. At the exit of the shredder, a highly entropic and difficult to separate mixture is obtained beyond the recovery of low quality steel, aluminium and copper, obtained from physical processes.

Since the electrical and electronic equipment concentrates most of the gold, silver, rare earths and elements of the platinum group, and whose thermodynamic rarity is therefore high, it is essential that these car parts be removed and treated adequately and in a way that is independent of conventional fragmentation processes. That is, these parts should be treated as WEEE (Waste Electrical and Electronic Equipment) and not as parts of the vehicle. As far as possible, it would be advisable to disassemble vehicles by the type of steel they contain. Second, in the fragmentation process itself, the parts of the vehicle could be crushed separately, depending on their composition and use, so that different scrap grades can be used for different grades of steels.

Naturally, an authorised treatment centre would say that these tasks, although desirable, are technically and economically unfeasible. The recyclers do not know the exact composition of the vehicles to properly separate the parts, there are no standard tools to disassemble the parts of the vehicle and even if there were, the disassembly times would be too long for the service to be profitable.

Therefore, it is essential to facilitate the disassembly of parts. For example, large parts such as the engine or gearbox are incorporated into the vehicle in less than

30 s on assembly lines, but their removal can take more than an hour. In addition, it is crucial to invest in eco-design measures that optimise the recycling of the parts, standardise the techniques and tools for disassembly and to ensure that authorised treatment centres have sufficient information on the composition and disassembly of each part, especially valuable ones.

Let us now look at the specific case of the vehicle analysed in Sect. 7.2. A total of 31 components with a high thermodynamic rarity had been selected for containing critical materials. It is precisely on these parts that eco-design measures should be focused, which can be broadly separated into four different categories (Ortego et al., 2018b):

1. Facilitating the disassembly of parts
2. Substituting critical metals with less critical ones
3. Retrofitting
4. Other approaches.

In addition to the thermodynamic rarity of the different metals that make up the vehicle parts identified as critical, information from the vehicle manufacturer itself was also taken into account in the study when formulating these eco-design proposals, in addition to opinions from recycling centres and personal interviews with interested parties.

An example of a component whose disassembly could be facilitated is the generator. Currently, the generator is located in the lower part of the engine. If it was placed on top, which could easily be done, the metals it contains—such as cobalt, nickel or tin, among others—could be more easily recovered. The same happens with the control panel (Fig. 8.9); it could be designed so that it could be easily detached from the dashboard, making it unnecessary to pre-remove the steering wheel and airbag.

For copper cables that are connected to the battery, they could also be disassembled at authorised vehicle treatment centres by removing the batteries. Finally, the disassembled assembly of the engine, gearbox and front axle would allow specific recycling processes to be used for the different elements present in the suspension arms (Nb, Mo, V), turbo (Nb, Cr, W), exhaust pipe (Pd, Ni, Zr), exhaust pipe temperature sensor (Pt, Ni, Cu), etc. This could be done easily as these components are designed to be easily and quickly placed on the assembly line at the time of automobile manufacturing, so carrying out the reverse disassembly process should not be a huge effort.

The second eco-design measure is focused on the substitution of some elements with others. It is to be remembered that within a vehicle there are many elements considered critical; therefore, alternatives to the use of these elements could be sought, replacing them with others whose criticality is lower, while maintaining their properties such as density, electrical conductivity, thermal conductivity, etc.

In the case of vehicles, the metals most used by weight in the analysed components, leaving aside iron and aluminium, are copper, gold and tantalum. Specifically, of the components identified as critical, copper appears in 30 of them, gold in 20 and tantalum in 19 (Ortego et al., 2018b).

Fig. 8.9 Dashboard of a Seat León III whose disassembly would lead to a greater recovery of the metals that compose it (Ortego et al., 2018b)

Copper is used primarily in cabling, and today copper cables are often replaced by aluminium cables. Considering that aluminium has 61% of the conductivity of copper, but 30% of the weight, it is a clear alternative to consider. An aluminium cable with a conductivity equal to that of the copper cables present in the vehicle would have 50% less weight but would be thicker.

In addition, to make aluminium cables conductive, elements such as iron, copper, magnesium or chrome are often added, making them less recyclable. Furthermore, in electronic contacts, the use of gold could be replaced by silver. Gold, with its high conductivity, corrosion resistance, high melting point and reflectivity, makes it widely used in electronics. However, due to their high cost, the parts are seldom made entirely of gold and those pieces, where necessary, are usually covered with a thin layer of gold. As an alternative, this coating could be made with silver, a cheaper and less critical material, considering that the latter is less resistant to corrosion, something that could be solved by adding nickel to the mixture.

Finally, the tantalum used in capacitors, completely lost in all metallurgical processes, could be replaced by ceramic capacitors. These ceramic capacitors have lower cost, smaller size, reliability and a longer service life. Thus, whenever possible, they should be used instead of tantalum capacitors. However, from an environmental point of view, ceramic capacitors require higher energy consumption for their manufacturing, as well as the use of nickel paste, so one would need to carefully analyse the risks and benefits of replacing some capacitors with others, not only from the point of view of the criticality of the elements used but also from the point of view of the effects on the environment.

We now come to the third eco-design approach: retrofitting. The fact that a vehicle reaches a recycling point does not mean that all its parts have reached the end of their useful life or that they have lost their functionality. In fact, the useful life of many of the components of the vehicle is longer than that of the vehicle itself, something well known to those dedicated to the recycling and sale of used auto parts. If the useful life of certain components is long enough, those same components can be reused in new or second-hand vehicles, which helps keep materials in use, reducing environmental impact and manufacturing costs. The demand for raw materials and critical metals when manufacturing new cars could be reduced, and the time that a component remains in the technosphere could be lengthened. However, this is something that must be done with certain precautions since there may be reluctance on the part of potential buyers regarding their use. It should be noted that this retrofitting or reconditioning process is already carried out for some large car components, even some that are not visible to the user, as is the case with industrial vehicles, whose engines may come from other already discarded vehicles. The components that could be ideally reused analysed are the dashboard, the lighting switches, the exhaust gas temperature sensor, the rear screen cleaner motor (Fig. 8.10), the rain sensor or the air quality sensor, among others. Reuse of some gearbox or engine parts could also be considered.

Other opportunities that may arise when eco-designing vehicle components are related to changes in the components or requirements of the car. For example, a few years ago, vehicles had built-in telephones, but nowadays many entertainment

Fig. 8.10 Rear screen cleaner motor of a Seat León III that could easily be reused in new vehicles (Ortego et al., 2018b)

systems can connect via Bluetooth to the user's own smartphone. Another example is replacing CD players with USB ports. Another solution would be centralisation of all the electrical units (on-board supply control unit, door unit, airbag unit, electronic control unit, etc.) in one, so that it could be easily disassembled in recycling centres and be sent to specific treatment centres, as is already the case with batteries or tires.

The four types of proposed measures can also be divided into two groups depending on who can and should carry them out. In the first case, the measures related to the substitution of some elements by others fall on the manufacturer. In other cases, such as the recovery and reuse of components, not only would manufacturers be involved in having to design the vehicles so that the parts can be extracted well but also the recycling centres and those responsible for their subsequent treatment.

The economic aspect is very relevant here since the costs associated with these measures and the benefits should be able to be distributed among the different parties involved. If we consider that the equivalent of €175 in critical and valuable metals is lost in an average vehicle, combined with the average life of a car (17 years) and the estimated sales of a model (150,000), losses could exceed 170 million euros (Ortego et al., 2018a). The question that remains to be answered is whether this economic loss justifies the implementation of measures related to eco-design that aim to improve the recovery of materials at the end of life of products. The answer can be found if we analyse the costs involved in the recycling process and, going further, in replacement costs avoided by preventing the extraction of virgin material from the Earth's crust. Let us see now how these can be evaluated.

8.6   Alternative Sources on Earth: Urban Mining

Reducing the consumption of natural resources is one of the keys to mitigating the effects of climate change in addition to reducing the waste generated by society. The most common solutions, as we have seen, fall on recycling to obtain secondary raw materials that can replace the primary ones. On the other hand, in the industry itself, a series of eco-design measures can be established to facilitate this recovery and reuse of certain critical metal. In the event that they cannot be recovered technically and economically, the last resort is to search for other metals that are less critical and can substitute them, maintaining the final properties of the products wherever possible.

There are certain pioneering alternatives that, in many cases, have not yet been extensively implemented or that are simply ideas on paper, which could potentially become viable sources of raw materials. Closely linked to recycling is urban mining, which consists of searching for raw materials not from conventional primary sources, such as mines, but from secondary sources, such as waste produced daily by a city, region or country.

As we have seen, a considerable part of electronics, construction, etc. waste ends up in manned or unmanned landfills, although in some cases, such as the automobile industry, the recycling and recovery rate is very high. The materials that make up the products that end there are often very valuable despite the small amount in weight they represent and are accumulated without the possibility of being recovered to give them a second or third life. In other words, these landfills could be seen as large human-made mines, where raw materials have been intentionally accumulated and which could well be used as an alternative source to mines for the extraction of many metals.

These metals often have higher concentrations in anthropogenic mines than in natural deposits. For example, in the case of gold, the mines that are exploited today contain on average between 4 and 10 g of gold per ton extracted; in the event that we had a landfill made up of smartphones, we would have a concentration of approximately 110 g per ton, and other studies estimate these amounts between 200 and 350 g of gold per ton of smartphones and computers (Hagelüken and Corti, 2010). Even in the case of gold mines with higher grades, such as the one in Kedrovka, Russia, whose average grade is 22 g per ton, we would not achieve such a high recovery rate. Similarly, up to 155 g of tantalum, another very rare metal, almost 51 kg of cobalt and more than 60 kg of copper could be recovered from one ton of smartphones. Another example is the project carried out in the old Penouta mine, in Spain, where it is estimated that there are a total of 12 Mt of mining waste, with an average concentration of 35 ppm of tantalum and 428 ppm of tin (Llorens González et al., 2020).

When we talk about urban mining, we are not only referring to the resources located in an urban environment but also to resources that are in the technosphere. In this way, different sources of material recovery can be established depending on where they are found, and we can separate six categories of materials: in use, hibernation, dissipation, landfill, slag and dumps (Johansson et al., 2013).

By size, the material in use represents the largest reservoir, representing around 50% of the material available in the technosphere. These materials in use could be strategic due to changes in the market that, despite their current use, may be used for other much more relevant or necessary applications. The materials in hibernation are accumulations of obsolete materials, that have reached the end of their useful life, that even today could be used as secondary sources: examples of this type of materials are old water supply systems, telecommunications networks, industrial structures, etc. Landfill mining, also known as ELFM (Enhanced Landfill Mining), refers to the extraction, processing, treatment and recovery of elements from material accumulations. Landfills and tailings might make up 10% of the technosphere's resources, while slags, hibernating materials, among others, account for less than 5% (Johansson et al., 2013).

Secondary mining, meanwhile, is the extraction and processing of rubble and slag, materials that have been initially discarded during the primary mining process. For example, in 2004, 2% of copper was produced from reprocessing of mine tailings (Graedel et al., 2004). The most complex source of recovery is the one that takes into account the dissipated materials, those that have dispersed again on land, sea or air, and whose recovery could almost be equivalent to mining on Thanatia, requiring prohibitive amounts of energy.

Urban mining is technically possible in all the types of environments described above, but in practice, there are very few places on the planet where viable initiatives

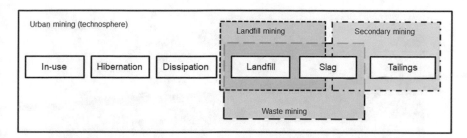

Fig. 8.11 Urban mining and its connection with different accumulations of materials in the technosphere. After Johansson et al. (2013)

are taking place, which are mainly focused on landfill mining and secondary mining (Fig. 8.11). We will now discuss pilot projects and experiences that have been carried out in relation to landfill mining.

In Austria, a study was carried out between 2016 and 2017 on the potential for recovery of different materials from waste deposits, specifically in the Halbenrain landfill, located 75 km southeast of Graz, consisting mainly of industrial-type waste. (García López et al., 2018). In the pilot project carried out, the material extracted from the landfill was taken to a biological and mechanical treatment plant with the aim of separating waste-derived fuels on one hand and metals on the other. Of the nearly 2,800 tons of waste that were processed, approximately 90 tons of ferrous metals were recovered. However, despite the success, the economic benefit obtained would not compensate the investment, because the total quantity is low compared with the current mines, although in the percentage of metal, it is greater than in the mine. Therefore, in addition to the extreme variability of the composition of landfills and the presence of toxic materials, it would not be economically profitable to carry out this recovery on a large scale. A similar study was carried out in the Styria region, also in Austria, analysing the potential of 10 landfills, with a theoretical amount of 5,154,700 tons of recoverable material, of which around 5% was waste with metallic content (Wolfsberger et al., 2013).

In Belgium, the Closing the Circle project (CtC) is one of the pioneering experiences in valuing the materials accumulated in landfills, specifically in one located in the municipality of Houthalen-Helchteren, in the northeast of the country, consisting mainly of waste from homes and industries (Quaghebeur et al., 2013). This deposit is operated by Remo Milieubeheer, from the Machiels Group, and the objective will be to convert the raw materials already used into resources, trying to make the final residue almost non-existent and mitigating CO_2 emissions. Thus, through a plasma installation, the waste deposited in Remo will be extracted and converted into energy and reusable raw materials. Similarly, it seeks to regenerate the land occupied by the landfill, so that it recovers its value and previous natural appearance. However, it is a very long-term project and the results will take time to see. In Spain, studies have also been carried out on the recovery potential of certain materials from landfills from a theoretical point of view, finding that the amount of ferrous and non-ferrous

Fig. 8.12 Debris from the demolition of a building: various materials that can be seen here could have a second life. *Source* Peakpx

metals that could be recovered is 2.9 and 0.5%, respectively (Puig Ventosa et al., 2014).

Another sector, in addition to the electronics sector, in which urban mining is booming is that of construction, where the generated waste can be easily used, after prior treatment, to construct new buildings (Arora et al., 2020) (Fig. 8.12). For example, from an old building with ten residential units, up to 1,500 tons of material can be recovered, including 70 tons of metals and 30 tons of plastics and wood (German Environment Agency, 2019).

As mentioned, secondary mining refers to the processing of slag heaps, slag and settling ponds that have been generated after primary mining, that is, it is the treatment of waste from the mining industry itself. In mining, it is common for certain metals to end up in the tailings due to their low concentration, due to the fact that, at the time of extraction, whether decades or centuries ago, their extraction was not profitable, had no known applications or the right technology for its recovery was not available. Currently, numerous mining companies are studying the existing tailings to analyse the feasibility of recovering the metals that have accumulated in them. For example, past processing and treatment methods of materials containing rare earths were not very efficient, so large amounts of these elements ended up in tailings. The concentrations of secondary deposits are sometimes so high that they can almost be considered mines. An example is the tailings of the Mountain Pass mine, in California, USA, where the concentration (between 3 and 5%) is so high that it is considered the second-largest rare earth deposit in the entire country, the first being the Mountain Pass mine itself. The same is true of waste from other processing plants spread around the world, such as that of La Rochelle, in France, or the tailings of Baotou, in China (Binnemans et al., 2015).

In the case of Spain, reprocessing of tailings and sludges is also being carried out, the most notable example being that of the company Strategic Minerals, where, since 2018, niobium and tantalum have been recovered from the waste from the Penouta mine, in Orense, which was exploited by Rumasa until 1985 to extract tungsten and tin (Strategic Minerals Spain, 2016). It is estimated that, during the old mine operation, around 15 million tons were left untreated. An additional factor is that during that time neither the demand for niobium and tantalum was very high nor was technology available to recover these elements. In addition to niobium and tantalum, other resources such as clays, quartz, feldspar and mica are also being obtained as by-products.

Although urban mining may seem like a promising solution, there are several associated problems. First, waste in landfills often appears mixed with large amounts of other materials such as plastics, wood, textiles, etc., which requires a pre-treatment stage that is not capable of competing with primary extraction. Second, the concentrations are often so low that the amount of energy required to extract the metals could be prohibitive. In the case of the elements that appear associated with other base metals such as copper, iron or zinc, their concentrations in this type of secondary sources are usually so low that they end up being at least comparable to those of mines or, in the last extreme, to Thanatia's. These concentrations would require much more energy for extraction than in conventional refining and beneficiation processes (as seen in Sect. 4.3). Similarly, in the case of urban mining, the metals appear mixed in such a way that their recovery is not always possible. For example, it is practically impossible to recover the lithium used in ceramic coatings. Therefore, despite the fact that urban mining may allow us to recover certain metals in the future, current demand will continue to be covered by primary mining extraction, as the rarity of the materials that could be recovered from secondary sources is not considered.

8.7 Alternative Sources Beyond Earth: Asteroid Mining

Another option that is being explored today as a possible source of certain metals to supply the growing demand is not on Earth, but in space. Although more than 20,000 near-Earth asteroids with a diameter greater than 100 m are known, as well as many smaller ones, the composition of just a few is likely to make them economically profitable. However, composition is not the only critical factor when analysing the viability of these asteroids: to be considered are also the grain size of the particles, the cohesion between them, etc., factors that could directly influence the mining process. These asteroids are not only studied for their mining potential but also for security reasons, no matter how remote that possibility may be—in fact, we know that some have historically impacted our planet. A recent example is a 17-m meteorite that exploded over Chelyabinsk, Russia, in February 2013, releasing energy 30 times greater than that of the Hiroshima nuclear bomb; fragments of different sizes were recovered nearby, including one weighing 650 kg (Fig. 8.13 and 8.14).

Fig. 8.13 Wake left by the passage of a meteorite over Russia on 15 February 2013, which exploded over the city of Chelyabinsk. Author: Alex Alishevskikh. CC BY-SA 2.0 Wikimedia Commons

Fig. 8.14 Fragments of different sizes recovered from the meteorite that exploded over Chelyabinsk (Russia) on 15 February 2013. Author: Alexander Sapozhnikov. CC BY-SA 3.0 Wikimedia Commons

The origin of these near-Earth asteroids is in gravitational interactions with Jupiter and Saturn, placing them at distances less than 1.3 astronomical units (195 million km) from the Sun and less than 45 million km from Earth (Galache et al., 2015). Today about 1,000 near-Earth asteroids are discovered each year and there are various projects, including NASA, that seek to characterise them.

The study of asteroids can be divided into three phases: 1) discovery of an asteroid of sufficiently large size, 2) remote characterisation by telescope and 3) characterisation in situ, which can, in turn, be divided into two parts, close and contact characterisation (Badescu, 2013). Each of these stages requires different research techniques and methods, becoming more expensive as the study progresses so that in each one of them the number of asteroids considered is less and less.

For an asteroid to be considered "discovered", a series of characteristics must be known about it, including its orbit, to understand when it passes close to Earth. When carrying out remote characterisation, spectroscopic and photometric techniques are used. By means of spectroscopy, the interaction between electromagnetic radiation and matter can be studied, and it is based on detecting whether said matter absorbs or emits electromagnetic radiation at certain specific wavelengths. In this way, information can be obtained about the asteroid's internal structure and temperature and also about the mineral composition on the surface. Photometry is used to determine the rotation and shape of the asteroid. Finally, in the case of the in situ characterisation of the asteroid, ships can be sent that stay close to the asteroid or that can even come into contact with it, combining different techniques such as X-ray diffraction, gamma-ray spectrometers, etc., to know more precisely the composition of the asteroid's surface.

Not all near-Earth asteroids are going to have identical composition, nor could they all serve as sources of rare metals. Depending on the composition, three categories of asteroids have been proposed: *silicaceous*, carbonaceous and metallic (Badescu, 2013). *Silicaceous* asteroids are made up, like chondrite type meteorites, of small spherical particles of different materials. Carbonaceous asteroids contain high levels of complex organic molecules and ice. Finally, metallic asteroids are for the most part made up of iron and nickel and can contain high concentrations of other heavy metals. Although we have this classification of asteroids, and a similar one for meteorites, it is practically impossible to establish links between certain types of meteorites and asteroids.

Due to the high cost of extraction that the metals present in asteroids will have only those elements that have a very high value in proportion to the mass of the asteroid will be interesting. For example, there are some estimates focused on asteroids that could contain enough elements of the platinum group (platinum, rhodium, osmium, iridium, palladium, etc.) so that their exploitation would be profitable at the prices that these elements have today. Assuming that there are those already mentioned 20,000 near-Earth asteroids with a diameter greater than 100 m, only around 4% of them could contain metals in non-negligible quantities. Of these, approximately half would have a concentration high enough to be profitable considering the extraction and transport costs. According to the calculations carried out, there could be a total of eight asteroids that could be viable sources of these elements (Badescu, 2013). Thus, an asteroid rich in these elements with a diameter of 500 m could contain up

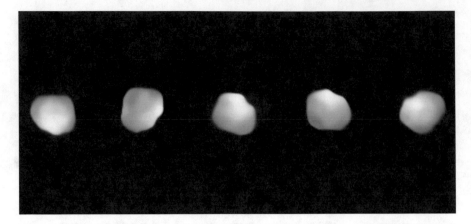

Fig. 8.15 Images of the Psyche asteroid taken during the HARISSA project, aimed at taking high-resolution images of different asteroids. Author: European Southern Observatory. CC BY 4.0. Wikimedia Commons

to 174 times the annual production of platinum (Lewicki et al., 2013). However, the problem is that we do not know what those 8 asteroids out of the 20,000 previously mentioned are, nor the energy cost that would have to be assumed to bring them to Earth.

An asteroid whose composition is known is Psyche, discovered in 1852 and measuring around 226 kms in diameter (Fig. 8.15). This asteroid is primarily made up of iron and nickel, among other precious metals, and is thought to be the rest of the nucleus of a primitive planet. However, leaving aside the economic value that the sale of the metals that compose it could have, and that some estimate in several quadrillion dollars, the real interest goes far beyond being a mere source of materials. It could provide valuable information on the nucleus of our own planet, of similar composition, and also on its formation.

Currently, there are several companies dedicated to research in different areas related to asteroid mining, among which stand out *Planetary Resources*, founded in 2009 by Eric C. Anderson and Peter H. Diamandis. Using a space telescope, their goal is to identify near-Earth asteroids that show the greatest potential for resources, both water and mineral. Planetary resources were acquired in 2018 by the company ConsenSys, which seeks to build a collaborative platform to diversify and decentralise space research.

As asteroid mining is one of the most abstract fields as a source of mineral resources, there are numerous studies that focus on economic viability, technological feasibility, legal aspects or even environmental aspects. How much would it cost to bring one to Earth? Who owns asteroids? Would they cause more or less environmental impact than mining on Earth?

Let us think for a moment about the implications of this. It has been calculated that the cost of sending a kilogram of material to space ranges between 10,000 and 20,000 dollars (Lewis, 1997). That is, sending something to space is very expensive, but transporting it down to Earth is not so expensive, in part because gravity plays in our favour. However, not only would we need to find an asteroid that could serve as a source, we would also have to think about how said asteroid could be processed and how we could transport the material to Earth.

In the case of those metals that are reasonably priced, space mining does not make much sense, since a much more inexpensive way could surely be found to extract them from terrestrial sources than to resort to space. In the case of metals with very high prices, such as platinum, perhaps the costs could offset these high costs. For example, a calculation made by Lewis (1997) on the near-Earth asteroid named Amun, only 2 km in diameter, estimates that its sale would generate more than 20,000 billion dollars of profit, among the iron, nickel, cobalt and elements of the platinum group present in it. This estimation assumes that it had a weight of 3 $\times$ 10^{10} tons and a standard composition of the metallic meteorite. On Earth, we can only extract materials from the first kilometres of the Earth's crust, but in the case of an asteroid, we could take full advantage of it. In any case, everything that has been carried out so far is pure speculation and we are in reality still far from being able to extract strategic materials from space.

We should not ignore the scientific interest that many of the asteroids have in explaining the origin of the solar system, planet Earth and life on it. Furthermore, even assuming that the exploitation of the materials that compose them could be carried out, the limitations that asteroid mining presents are multiple. First, although we may know the external composition of an asteroid, this does not imply that the entire asteroid in its entirety is homogeneous, that is, the calculations mentioned above may not be correct and the Amun "value" may be much lower than calculated. We also do not know if the techniques applied in mining today could be extrapolated to asteroid mining, where gravity does not exist or where there is no presence of water, both fundamental agents in mining on Earth, or if it would be easier to bring asteroids to Earth or to introduce them into the lunar orbit, instead of trying to process them in space, with the consequent risks that this entails. Therefore, although asteroid mining could solve the supply problems of some elements, such as elements of the platinum group or rare earths, today, and probably for decades to come, it is more a dream than a reality.

References

Allwood, J. M., Laursen, S. E., de Rodríguez, C. M., Bocken, N. M. (2006). Well dressed? The present and future sustainability of clothing and textiles in the United Kingdom. Available at: https://www.ifm.eng.cam.ac.uk/uploads/Resources/Other_Reports/UK_textiles.pdf.

Arora, M., Taspall, F., Cheah, L., Silva, A. (2020). Buildings and the circular economy: Estimating urban mining, recovery and reuse potential of building components. *Resources, Conservation and Recycling, 154.* https://doi.org/10.1016/j.resconrec.2019.104581.

Ayres, R. U. (1994). Industrial metabolism; Theory and policy. In B. R. Allenby & D. J. Richards (Eds.), *The greening of Industrial ecosystems* (pp. 23–37). National Academy Press.

Badescu, V. (2013). *Asteroids.* Springer.

Benyus, J. M. (2002). *Biomimicry: Innovation inspired by nature.* Harper Perennial.

Binnemans, K., Jones, P. T., Blanpain, B., Van Gerven, T., & Pontikes, Y. (2015). Towards zero-waste valorisation of rare-earth-containing industrialprocess residues: A critical review. *Journal of Cleaner Production, 99*, 17–38.

BIO by Deloitte. (2015). Study on data for a raw material system analysis: Roadmap and test of the fully operational MSA for raw materials. Prepared for the European Commission, DG GROW.

Bouyer, E. (2019). Substitution strategies guide for R&D&I. Solutions for Critical Raw Materials—A European Expert Network (SCRREEN). WP5. D5.2.

Chertow, M. R. (2007). "Uncovering" industrial symbiosis. *Journal of Industrial Ecology, 11*, 11–30.

Cramer, J. (1997). Towards innovative, more eco-efficient product design strategies. *Journal of Sustainable Product Design, 1*, 7–16.

CRM InnoNet. (2013). Critical raw materials substitution profiles. Critical Raw Materials Innovation Network.

CRM InnoNet. (2015). Roadmaps for the substitution of critical raw materials. Critical Raw Materials Innovation Network.

EMF. (2015). Towards a circular economy: Business rationale for an accelerated transition. Ellen MacArthur Foundations, Isle of Wight, United Kingdom.

ERECON. (2015). Strengthening the European rare earths supply-chain: Challenges and policy options. European Rare Earths Competency Network.

European Commission. (2018). Report on critical raw materials and the circular economy. https://doi.org/10.2873/331561.

European Commission. (2017). Study on the review of the list of critical raw materials. Critical raw materials factsheets.

European Commission. (2013). Strategic implementation plan for the European innovation partnership on raw materials. Part II: Priority areas, actions areas and actions.

Frosch, R. A., & Gallopoulos, N. E. (1989). Strategies for manufacturing. *Scientific American, 261*, 144–152.

Galache, J. L., Beeson, C. L., McLeod, K. K., & Elvis, M. (2015). The need for speed in Near-Earth asteroid characterization. *Planetary and Space Science, 111*, 155–166.

García López, C., Küppers, B., Clausen, A., Pretz, T. (2018). Landfill mining: A case study regarding sampling, processing and characterization of excavated waste from an austrian landfill. *Detritus, 2*, 29–45. https://doi.org/10.31025/2611-4135/2018.13664.

Gerber, J. (2007). Strategy towards the red list from a business perspective. In R. W., Scholz, & D. Lang (Eds.), *From availability to accessibility—Insights into the results of an expert workshop on "mineral raw material scarcity".* Davos, Switzerland.

German Environment Agency. (2019). Urban mining: resource conservation in the anthropocene. Available at: https://www.umweltbundesamt.de/sites/default/files/medien/479/publikationen/uba_urbanmining_en_2019.pdf.

Graedel, T. E., van Beers, D., Bertram, M., Fuse, K., Gordon, R. B., Gritsinin, A., Kapur, A., Klee, R., Lifset, R., Memon, L., Rechberger, H., Spatari, S., & Vexler, D. (2004). The multilevel cycle of anthropogenic copper. *Environmental Science and Technology, 38*, 1253–1261.

Hagelüken, C., & Corti, C. W. (2010). Recycling of gold from electronics: Cost-effective use through "Design for Recycling." *Gold Bulletin, 43*, 209–220.

Iglesias-Émbil, M., Valero, A., Ortego, A., Villacampa, M., Vilaró, J., & Villalba, G. (2020). Raw material use in a battery electric car—A thermodynamic rarity assessment. *Resources, Conservation and Recycling, 158*. https://doi.org/10.1016/J.RESCONREC.2020.104820.

Jiménez Herrero, L. M., & Pérez Lagüela, E. (coord) (2019). La economía circular en el paradigma de la sostenibilidad. In Economía Circular-Espiral: Transición Hacia Un Metabolismo Económico Cerrado. Asociación para la Sostenibilidad y el Progreso de las Sociedades (ASYPS) y Ecobook.

Johansson, N., Krook, J., Eklund, M., & Berglund, B. (2013). An integrated review of concepts and initiatives for mining the technosphere: Towards a new taxonomy. *Journal of Cleaner Production, 55*, 35–44.

Kirchherr, J., Reike, D., Hekkert, M. (2017). Conceptualizing the circular economy: An analysis of 114 definitions. *Resources, Conservation and Recycling, 127*, 221–232. https://doi.org/10.1016/j.resconrec.2017.09.005.

Lee, M., Hong, J., Lopez, J., Sun, Y., Feng, D., Lim, K., Chueh, W. C., Toney, M. F., Cui, Y., & Bao, Z. (2017). High-performance sodium–organic battery by realizing four-sodium storage in disodium rhodizonate. *Nature Energy, 2*, 861–868. https://doi.org/10.1038/s41560-017-0014-y.

Lewicki, C., Diamandis, P., Anderson, E., Voorhees, C., & Mycroft, F. (2013). Planetary resources—The asteroid mining company. *New Space, 1*, 105–108.

Lewis, J. S. (1997). Mining the sky: Untold riches from the asteroids, comets, and planets. Helix Book.

Lim, X. (2016). New high-entropy alloys metal mixology. *Nature, 533*, 306–307.

Llorens González, T., Mateos Aquilino, V., García Polonio, F. (2020). La mina de Penouta, minería sostenible para el abastecimiento de Ta & Nb en Europa. Semin. la Soc. española Mineral. 14.

McDonough, W., & Braungart, M. (2002). *Cradle to cradle*. North Point Press.

Mizuno, F., Yada, C., Iba, H. (2014). Solid-state lithium-ion batteries for electric vehicles. In *Lithium-Ion Batteries*. pp. 273–291.

Ortego, A., Valero, A., Valero, A., & Iglesias-Embil, M. (2018a). Downcycling in automobile recycling process: A thermodynamic assessment. Resour. *Conservation & Recycling, 136*, 24–32.

Ortego, A., Valero, A., Valero, A., & Iglesias, M. (2018b). Toward material efficient vehicles : Ecodesign recommendations based on metal sustainability assessments. *SAE International Journal of Materials and Manufacturing, 11*, 213–227. https://doi.org/10.4271/05-11-03-0021.

Pauli, G. (2010). The blue economy. 10 Years, 100 Innovations, 10 Million Jobs. Paradigm Publishers, Taos, New Mexico.

Pavel, C., Marmier, A., Alves Dias, P., Blagoeva, D., Tzimas, E., Schuler, D., Schleicher, T., Jenseit, W., Degreif, S., Buchert, M. (2016). Substitution of critical raw materials in low-carbon technologies: lighting, wind turbines and electric vehicles. EUR 28152 EN. Publications Office of the European Union; 2016. JRC103284.

Potting, J., Hekkert, M., Worrell, E., Hanemaaijer, A. (2017). Circular economy: measuring innovation in the product chain. PBL Netherlands Environmental Assessment Agency. Available at: https://www.pbl.nl/sites/default/files/downloads/pbl-2016-circular-economy-measuring-innovation-in-product-chains-2544.pdf.

Puig Ventosa, I., Calaf Forn, M., Jofra Sora, M. (2014). Urban mining extracting resources from landfill sites. *Seguros y Medio Ambientales, 34*.

Quaghebeur, M., Laenen, B., Geysen, D., Nielsen, P., Pontikes, Y., Gerven, T. V., & Spooren, J. (2013). Characterization of landfilled materials: Screening of the enhanced landfill mining potential. *Journal of Cleaner Production, 55*, 72–83.

Reuter, M. A., Hudson, C., van Schaik, A., Heiskanen, K., Meskers, C., Hagelüken, C. (2013). Metal recycling: Opportunities, limits, infrastructure. Available at: https://www.resourcepanel.org/reports/metal-recycling.

Schmidt-Bleek, F. (1994). *Wieviel Umwelt braucht der Mensch?–MIPS-Das Mass fur ekologisches Wirtschaften*. Birkhauser Verlag.

Strategic Minerals Spain. (2016). The Penouta project: Strategic and sustainable mining. Available at: https://www.phytosudoe.eu/wp-content/uploads/2016/11/10_Strategic-Minerals_Penouta-Project_PhytoSUDOE-workshop-2017.pdf.

Talens Peiró, L., & Castro Girón, A. (2020). Examining the feasibility of the urban mining of hard disk drives. *Journal of Cleaner Production, 248*. https://doi.org/10.1016/J.JCLEPRO.2019.119216.

UNEP (2011). Recycling Rates of Metals – A Status Report. A Report of the Working Group on Global Metal Flows to the International Resource Panel. In: Graedel, T. E., Allwood, J., Birat, J.-P., Reck, B. K., Sibley, S. F., Sonnemann, G., Buchert, M., Hagelüken, C. Available at: https://www.resourcepanel.org/reports/recycling-rates-metals.

Valero, A., Valero, A., Calvo, G., & Ortego, A. (2018). Material bottlenecks in the future development of green technologies. *Renewable and Sustainable Energy Reviews, 93*, 178–200.

Valero, Antonio, Valero, Alicia (2014). Thanatia: The destiny of the Earth's mineral resources: A thermodynamic cradle-to-cradle assessment. World Scientific Publishing Company.

Verhoef, E. V., Dijkema, G. P. J., & Reuter, M. A. (2004). Process knowledge, system dynamics, and metal ecology. *Journal of Industrial Ecology, 8*, 23–43.

Weizsacker, E., Lovins, A. B., Lovins, L. H. (1998). Factor four: Doubling wealth, halving resource use—A report to the club of Rome. Routledge.

Wolfsberger, T., Höllen, D., Sarc, R., Pomberger, R., Budischowsky, A., Himmel, W., Mitterwallner, J. (2013). Landfill mining Austria—Pilot reegion Styria. In *ISWA Conference*. Serbia.

Chapter 9
Epilogue: For a New Humanism that Cares About the Future of the Planet

Abstract It is broadly true that matter here on Earth, like energy everywhere, is conserved but degrades. If the energy of a system degrades until it reaches equilibrium with its environment, so also does the Earth's stock of economically valuable non-renewable materials of various kinds. However, there is a big difference, inasmuch as the Sun renews the energy of our planet every day but does not mend the degradation and dispersion of Earthly materials. That dispersion has now been exponentially accelerated by human agency, and so the Earth is tending swiftly now toward becoming a degraded planet which we called Thanatia—a doom which would entail the collapse of our civilisation. Market economists forgot this simple message of physics: Every economic benefit has an associated natural cost, which purveyors of market economics wish to ignore systematically; the value of the planet to Humankind is depreciating, yet its amortisation is ignored in our economists' accounts. To avoid the doom of Thanatia would depend on wise human agency. Technological solutions may be feasible, but only if technology is reoriented. And, it is not only the national economies that need to be changed, but also social perceptions. People need to adopt big-time, anti-entropic activities of separation (of different kinds of waste) and replenishment. While consumable energy can be obtained from various renewable or non-renewable sources, chemical elements cannot be transmuted into each other, and therefore, economically essential materials are often troublesome and sometimes impossible to replace with each other. In a high-tech economy depending on supplies of, say, 50 essential resources, economic collapse could be triggered by supply blockage of any one of those 50. A mature vision of our planet is needed, valuing precaution and valuing the experience of other, more conservative cultures, and admitting our responsibility for the finite future of the planet. A Universal Constitution is needed to ensure a modicum of balance between regard for civilisation and regard for the Earth's capacities to support civilisation. Such a Constitution would be based on a new kind of Humanism. This is the subject of this epilogue.

9.1   Reorienting Science and Technology of Materials

9.1.1   Materials, a Global Responsibility

Human fascination with life tends to forget its geological basis. Minerals and rocks are used to build cities and infrastructures, provide fertilisers for agriculture and supply fossil fuels that drive more than 87% of the world's energy. Today, they are also used on a mass scale to sustain the digital economy, which relies, above all, on scarce minerals whose depletion rate has skyrocketed, putting economic development itself at risk in less than a generation. There is no room for unlimited desires on a limited planet.

The solar engine naturally drives the chemical elements that constitute the living things, the atmosphere and the hydrosphere. In a sustainable world, all material cycles must be closed. This should be, of course, the case for the 25 life-supporting elements, including carbon, oxygen, nitrogen, phosphorus, calcium, sulphur and a few more trace elements. Yet also for those with a crustal origin primarily presented in a solid-state and whose natural circularity is limited even in geological eras.

Sun's energy seems insufficient to recycle these elements, and humankind is dissipating the planet's mineral wealth in a few generations. The extra-terrestrial energy coming from the Sun is approximately 2 cal per centimetre every minute. This solar constant is an immense energy source that daily moves the entire atmosphere, constantly circulates the hydrological cycle, warms the biosphere and stimulates plants' chlorophyll function. Unfortunately, it is not a concentrated source, and therefore, to melt, reduce, react, move, heat, cool, separate, etc., i.e., to carry out all economic activities, we need the energy intensity provided by fossil fuels. This means that human civilisation is developing on the margins of biomaterials. Today, mankind uses almost all the elements of the periodic table, including the radioactive ones, without considering the rarity of minerals.

Humanity cannot rely solely on extraction because demand for materials is growing exponentially. The throwaway culture is now unsustainable, making recycling, reuse and reducing consumerism all the more important. However, society is far from closing even the first cycle. We are limited to a bleak future unless drastic and unlikely changes are implemented.

In recent decades, problems that did not even exist or were not a priority before have emerged. Among them, the recycling of scarce materials. Yet there was no clear evidence of a rapid loss of the planet's mineral wealth, which requires fair compensation, particularly for countries exporting rare raw materials.

Added to this is the proliferation of thousands of new artificial chemical products impossible to be broken down by natural micro-organisms. These include the microplastics polluting the oceans and entering the food chain.

Materials are a problem of global responsibility that the Sun cannot solve on its own. It is imperative that all nations systematically account for the extraction and depletion of raw materials. In other words, global accounting of ongoing depletion and degradation should place nature at the same level as the global economy.

Such global accounting would promote new policy decisions such as new global tax structures to conserve natural resources.

9.1.2 Are There Technological Solutions?

Technological solutions do exist. Throughout the book, we have seen that we will have to make a solid commitment for dematerialisation, the substitution of critical materials for others that are more abundant or better bio-based, the extension of product lifetimes and reuse and recycling. Accordingly, it will be essential to design products with their end of life in mind, making them robust, modular and easily disassembled. This will enable to drastically reduce waste, reuse and repair products, giving them a second, third, fourth life, and recover valuable materials that are subsequently introduced into the production system. In this sense, we will have to learn from nature, which does not produce waste and lives and regenerates exclusively from the action of the Sun.

To that end, a reorientation of scientific and technological interests is crucial. Researchers are very interested in producing new materials with extraordinary properties, such as superalloys, nanostructures and organometallic substances. However, very few focus on separation techniques, such as demixing, partitioning, decontamination or the recovery of basic elements. Recovery is neither scientifically attractive nor economically profitable when creating new materials and bringing them to market. Just as no pharmaceutical product can enter the market until its side effects have been assessed, no new material/product should be placed on the market until its expected life cycle has been evaluated. In a finite material world, innovations in separation deserve the same recognition as the creation of new materials. This new science of separation needs thermodynamics to optimise processes.

That said, and knowing the limitations imposed by the second law of thermodynamics regarding the impossibility of completely closing cycles, the main measure to be adopted must be to reduce consumption, opening the way to new economic models that encourage "use" rather than "possession". An economy of services where companies do not sell the product but its function. In this way, products will be more robust and at the end of their life, producers will be responsible for recovering their valuable materials and reintroducing them into the system more efficiently.

On the supply side, if demand continues to grow, we will never be able to give up mining. This will make us face several contradictions, such as the "Nimby" effect (Not in my backyard). We do not want extractive activity nearby because of the impacts it generates. We prefer to relegate it to third countries, in many cases with no or low environmental and social standards. However, we do not give up the right to constantly renew technological products, which require mining activity for their manufacture. Reducing foreign dependence, which is a priority for many governments, implies a commitment to mining in one's own territory and probably opening or reopening new mines, which will most certainly create social rejection. In this sense, mining must be environmentally and socially sustainable in my backyard and in the backyard of my neighbours. The mineral wealth, which is a natural heritage of those who live today and those who will be born, must be fairly valued, not only considering the costs of extraction today but also those that future generations will have to face when mineral deposits become exhausted. This is the only way to create a true sense of conservation.

Avoiding dependence on fossil fuels will mean accepting dependence on materials, some of them with significant supply risks. Without materials, there is no energy, but without energy, there are no materials either! It is, therefore, necessary to consider the dialogue, or better trialogue energy—materials—environment, because the solutions will not be one-dimensional but multi-dimensional and complex, especially when the serious social problems associated with mining come into play.

9.2   For a New Humanism

9.2.1   A Quick Overview of the State of the Planet in the Twenty-First Century

The twentieth century was a century of mutual destruction with two world wars, colonial wars, civil wars, genocides and other atrocities. With the demise of the communist bloc and the triumph of liberal democracy, Francis Fukuyama proclaimed "the end of history" in 1992. The market and globalisation were elevated to the status of substitutes for bloody wars. Yet nothing could be farther removed from the truth. The trade wars that heralded prosperity turned into wars against nature and, in the process, local populations were overexploited, creating huge global inequalities. The fact that on a finite planet, there is no room for infinite ambitions and desires were never considered. Today, almost every corner of the Earth has been explored and will likely very soon be exploited. If the current trend continues, neither the poles, nor the oceans or the seabed, nor the tropical belts or the boreal forests, nor the fertile soils or the depths of the crust, nor the atmosphere or the stratosphere will be spared from being devastated.

We think that if we cut our CO_2 emissions by replacing fossil fuels with renewables, we will have solved climate change. However, the historical emissions released are already seriously affecting the environment. For example, ocean acidification is wiping out coral reefs, affecting the oceans' ability to retain CO_2, and the melting of the poles and permafrost in Canada, Alaska and Siberia are releasing trapped methane. Both effects will induce further climate change, even if the emission of greenhouse gases completely stops.

According to Johan Rockström and colleagues at Stockholm University (Rockström et al., 2009a, b), there are even more pressing problems that go unnoticed. For example, the excess fertilisers used in today's agriculture ultimately reach the oceans, altering the biogeochemical cycles of phosphorus and nitrogen. In the case of phosphorus, this leads to an irreversible loss of the planet's mineral capital, which threatens the future of human nutrition, as it is a vital and irreplaceable chemical element for all living beings. Further examples include the availability of fresh water, which causes climate refugees; the destruction of the stratospheric ozone layer, diminishing our protection against dangerous UVB rays; the loss of biodiversity that may result in, among others, the emergence of future pandemics; pollution with organic chemicals, heavy metals, micro-plastics or nuclear waste. The list is endless, and new threats appear every year. Worse still, some of them are irreversible. The Anthropocene has been unleashed.

Despite this, our blind trust in technology and some "greenwashing" legislation will encourage us to continue depredating while defending ourselves and adapting to increasingly violent "natural" reactions. We would rather listen to those who argue that the planet is full of opportunities and resources than to those who argue for a more balanced approach to Nature. Deep down, we are unwilling to accept moderation in consumption when it comes down to the personal level. We all need a little bit more, which we think is irrelevant, but it is not irrelevant across the whole of humanity.

At most, we accept that greenhouse gas emissions seriously affect climate change. We also consider that the uncontrolled production of chemicals should be reduced. In other words, we accept to pay more for decontaminating activities, essentially of hygienic and sanitary nature. These are end-of-pipe solutions, but they are by no means a truce with nature itself. Real solutions would bring about a severe restriction in our own consumption that would lead to an eco-social revolution.

In the decades to come, the world will continue to struggle with short-term shortages, occasionally generated by dramatic situations such as wars, natural catastrophes and accidents, as well as other economic, social and political problems, which will arguably interrupt global supply chains. In any case, and regardless of the cause, it will lead to an ever-increasing price rise of raw materials, which, although fluctuating, could become permanent, threatening the current status quo.

In the more or less long term for economics, and in the very short term for geology, mineral depletion will be seen as a problem when shortages of some minerals become apparent, at which point it may be too late to react. There are many examples of human-caused biological extinctions. Museums are full of stuffed animals, drawings and sketches of creatures that no longer exist. There will undoubtedly be "extinctions" in geodiversity. Our generation will not care better about mineral resources than biological ones or conserve them for future uses. This is a critical issue for the sustainability of the planet and of life.

9.2.2 Thanatia and Exponential Behaviour

The concept of Thanatia transcends mere scientific calculation, so it will not be readily accepted. Implicitly, it suggests that beyond the Anthropocene, the civilisation we have developed over these few thousand years can no longer exist. We have made a virtue out of greed, simplifying and fragmenting nature as a source of resources that can be exchanged for money. And this, to the extent that we only pay attention to GDP growth, and no country accounts for the natural deterioration that this entails. We consider socially acceptable growth of, for example, 2.5% per year without assuming that this implies an exponential decrease in nature of the same order of magnitude. In fact, if we carry on like this, we will have consumed what humanity consumed in its entire history in one generation, and in 25 years, we will consume twice as much as we do today. Nobody likes bad news. Even if the collapse was accepted as inevitable, it would arguably lead to a state of widespread destructive despair, what Paul Raskin of the Tellus Institute calls barbarism.

We have explained in the book that Thanatia is a quantitative model that allows assessing the speed with which our society is approaching it. Yet, it is not only the message of collapse that it provides. Two more messages derive from the second law of thermodynamics as applied to the planet. The first is that entropy behaves exponentially. All its formulas are logarithmic, never linear. In more intelligible terms, it means that more and more effort is needed to obtain smaller and smaller results, and in the limit, the efforts would be infinite with zero results. This applies equally to progressively more meticulous decontamination, as it does to increasingly greedy resource extraction. Indeed, both decontamination and mineral extraction have their limits imposed by economics (law of diminishing returns) and ultimately physics, which in thermodynamics is called "pinch points".

Although it may seem obvious, every finite system has its end. Thus, "any action that causes exponential effects in the components of a finite system leads to the collapse of the system itself, and this is independent of the indicator used to characterise the action." In other words, Thanatia does not need to be limited to the analysis of mineral extraction but to any human activity on nature with exponential behaviour. To simplify the message, we would say that entropy describes the exponential character of degradation. Therefore, any exponential degradation phenomenon can be characterised by entropy.

The second message is that of multiple limits. Given the complexity of nature, the lack or excess of a single resource puts at risk the system as a whole. It is known as the law of the minimum or Liebig's law: it states that "the evolutionary process of a system is not controlled by the total available resources, but by the scarcest resource." Since we do not know the complexity of life, we cannot predict what the Achilles heel of our own destruction will be.

In other words, in a very short geological time, the war against nature will be lost by our civilisation, because the availability of resources will become successively scarcer, and this will force greater consumption of energy, materials, and thus greater emissions, waste and degradation. It will also lead to the accelerated exclusion of more and more marginalised people and to global disorder through glaring inequalities. By fragmenting nature into resources, for the sole purpose of being consumed, we ignore the limits by which the web of life on Earth may collapse. This is not a prediction, but a consequence of the second law of thermodynamics. That said, thermodynamics cannot predict how long it will take, because such time to collapse depends more on human beings than on physics.

9.2.3 Youngsters or Mature?

Perhaps, the concept of Thanatia will never be popular. Humankind views death with horror, not as a natural process. As a tree grows, some branches die and others sprout again thanks to the work of the former. If there was no such replacement, the tree would die. Degradation is always pervasive.

The concept of Thanatia must give rise to a new humanism. Kenneth Boulding (Boulding, 1966) suggested the reflection that the Earth is a single spaceship, around which, even very far away, life is absent and presumably uninhabitable. If we destroy the ship, in several generations, our civilisation will have no future. We are treating the planet as if we were cowboys, i.e., people who, on the vast North American plains of the Midwest, were able to prey on their natural and human environment without awareness of limits.

By analogy, an adolescent vision of the planet prevails today in which there is no assumption of limits and no awareness of a future beyond our own lifetime. Short-termism and narrow-minded actions predominate in the concerns of most of today's leaders.

We, therefore, claim that the relationship of human beings with the planet should be one of maturity, not adolescence, with a responsible view of the eventuality of their finitude. The concept of maturity is characterised not only by greater experience, caution and wisdom but also by a feeling of being more or less close to death, and that idea can also be extended to the planet. Yet there is a difference: there will not be an end of the planet, but an end of our capacity to depredate, and with it a possible end of our civilisation.

From this perspective, it would be a matter of managing time backwards. Priorities change; we have to concentrate on what is essential. In the time that is left, we have to ask ourselves what we will do in the near future. We have to organise ourselves to make the remaining life as joyful and, above all, as long and healthy as possible. This mindset of respect, recognition and dialogue with nature is absolutely fundamental. It is the basis of a new humanism. We need to contemplate the planet from a vision of maturity.

9.2.4 Thanatia and the Backwards Vision of the Future

When future projections are designed, they are always presented in a variety of scenarios. The typical BAU (business as usual) is frequently the benchmark, and then other more or less optimistic scenarios are analysed. This forward-looking (forecasting) approach has been applied to energy, water, climate, fisheries, population, etc. Its results predict crises, conflicts or collapses of an economic, political, health or climatic nature that threaten our future. Yet there is always at least one scenario in which we could be saved.

Thanatia's message turns prognostic thinking on the use of natural resources upside down. Instead of moving from today to a defined temporal future, Thanatia's thinking suggests time to run backwards. Thanatia's message can be used to rethink the human–nature interaction. It would be a kind of evolutionary game theory, where solar action can regenerate some damage to the biosphere in human lifetime intervals, but most of the damage to the biogeosphere, i.e., the atmosphere, hydrosphere and crust, will only regenerate over long, even geological, times. Nature has the capacity for spontaneous regeneration with the Sun's power, but it does so at its own pace. Humankind, in turn, has intelligence, speed of action and artefacts to plunder its resources. If the speed of the planet's degradation exceeds its regenerative capacity, planetary entropy will grow and the Earth will dangerously approach Thanatia.

However, natural regenerative capacity is not one-dimensional; each dimension has its own speed, depending on geobiological and sociological circumstances. Therefore, the only way to avoid mutual collapse is to use human intelligence to accelerate the natural regenerative capacity or replace what has been destroyed, especially in those dimensions with a longer natural lifespan. However, human interests may hold back replenishment, either out of self-interest or because of its high cost. In the meantime, nature inexorably continues to accelerate towards degradation.

For example, regenerative agriculture should be promoted, which considers the replenishment of fertile soil as necessary as harvesting. This requires feeding the microbiota by replenishing the nutrients that have made the soil fertile. This has economic and social costs, but the sooner soils are replenished, the longer it will be possible to feed the current and future population.

Another example is the waste of rare chemical elements used profusely in electronic equipment. As we have seen in the book, a conventional vehicle contains more than 50 elements, of which more than 30 are scarce. A vehicle also has about 30 different plastic types, most of them non-recyclable and constituting a serious environmental problem. A technological breakthrough in the substitution of critical raw materials for less rare ones or a deep circular economy where these scarce materials are reused over and over again would prevent the depredation of the planet's most remote places.

Beyond an energy transition, a material and ecological transition of the planet is urgently needed to repair the damage done so far to nature. Continuing with current consumption is unsustainable, because we are doomed to an "every man for himself" type of barbarism. We must build a global partnership based on responsible people and institutions that will lead us to a stationary anti-entropic system, i.e., one that avoids exponential behaviour and systematically replenishes the damage caused throughout humankind's history on our single spaceship.

We need a system that allows us to assess our behaviour and the sustainability strategies of each decision-making level, with standard environmental cost indicators, suitable accounting systems, and the required processes to quantify progress, identify needs and interpret decisions.

However, an ecological transition without incorporating a profound change in the economy's conception is epistemologically unfeasible. The social attitude towards nature, which must be taken into account and accounted for, must be changed.

We cannot remain indifferent to the evidence of limits. We must take on board in all facets of our relationship with nature, the need to replace the damage caused. If this is not feasible, we must stop extraction radically until alternative ways can be found. If we are certain of the end, we can count backwards. This technique is called backcasting. It consists of imagining a possible future and putting in place the means to achieve it, in our case to avoid it or extend it. This planet needs a strategic plan.

9.2.5 The Need for a Strategic Plan for the Planet

The planet cannot wander aimlessly without knowing where resource waste and the fight against the very nature that hosts us is leading us. Although we do not have reliable accounts, today we already know the state of degradation. We even know the desired state of the planet, but the big question is how to get from the current situation to the desired one.

There should be neither winners nor losers in this planet's future, but peaceful coexistence between our civilisation and nature. Therefore, the overall objective must be to restore the damage, thereby reaching a steady coexistence state. To this end, it would be necessary to define the actions to be undertaken and the allocation of financial, human and logistical resources to achieve it, considering a planetary perspective without borders. At the same time, we need an accounting of the deterioration and replenishment achieved year after year, with periodic feedback analyses to correct mistakes and record progress. This accounting of damage and replenishment should be arguably based on Thanatia's model, which, using the second law of thermodynamics, would periodically measure homogeneously for all impacts the loss or gain, creating an accounting procedure similar to that used in traditional economic accounting. The essential indicators would be space, time, energy and the entropic (or exergetic) state of each natural dimension analysed. The universal objective would be zero damage, compensating degradations with replenishments, both measured by entropic/exergy indicators.

It should be a global effort in which all local/national projects would contribute with the same measurement units to assess the cost–benefit of their activities for nature. As feedback, the compensation activities that should be carried out and shared should be defined. Nature has no borders.

The global organisation should articulate, with a planetary perspective, global and local objectives, immediate and longer term objectives, defining a transparent, consensual and fair strategy. The model should be replicated at each and every decision-making level, in such a way that countries and their regions carry out the same standardised accounting, while maintaining the principles of subsidiarity (what can be done at a lower level should not be done at a higher one) and solidarity so that no one loses the perspective that the task is universal and long-lasting.

Clearly, a change in thinking and culture is needed. More and more researchers are advocating the need for a Universal Constitution, in which the responsibility of all human beings for the planet and its future would be enshrined. In this Universal

Constitution, the rules for global governance would be established, and these, in turn, would be articulated with national objectives. Such a cultural change is necessary to learn from mistakes, adapt and accelerate the processes of change. Of course, it is also necessary to allocate the resources and tasks that each decision-making unit should carry out.

In short, there is already a reasonably broad understanding of the planetary degradation that urgently needs to be addressed. A multidisciplinary group of concerned and knowledgeable people could develop a viable strategy, defining the goals and resources needed to replenish the planet. Such an authority would then involve all nations of the world in the process, building on the previous consensus reached at successive Conference of the Parties (COP). This is a crucial step because, without this involvement, nothing would be achieved. Similarly, it would be necessary to maintain solidarity structures to overcome the drawbacks and risks of failure that would indeed take place.

This is in itself a daunting task that will take a long time, beyond several generations. However, we believe that only with a global and comprehensive strategy could we achieve the planet–civilisation balance required for a long and dignified life of the human being, preventing Gaia from ever degrading to Thanatia, as long as the Sun keeps shining on us.

The arrow of time is not as harmful as it seems. We can say that humankind grows in knowledge and wisdom. If the increase in wisdom outpaces the further and irreversible deterioration of the planet, we will reach sustainable conditions with the Earth.

9.2.6 A New Humanism

Achieving any kind of sustainability means balancing opposites by seeking their complementarity rather than their confrontation. Balancing means, for example, compensating a nature fragmented in resources with an economy that replenishes and repairs the natural deterioration induced in order for both to survive. The search for balance between diversities gives meaning to the evolution of human beings on this planet. Therefore, it is key to achieve a balance between many opposing systems, such as those of gender, cultures, wealth, power, ideologies, and above all, the human–biosphere couple. Each couple is multidimensional, so simple solutions are not valid.

Our western culture has made out of difference the basis of confrontation and exclusion. The arrogance of always wanting to be right because of short-term success, greed and selfishness have subjugated nature, extinguishing its biodiversity and exploiting its geobiological wealth to the point of exhaustion. We have also subjugated people, destroying their cultures and taking over their territories.

This behaviour is accelerating. But this world is finite in resources and the diversity of different ways of life is gradually giving way. The result is a uniform planet, dwarfed and weakened by the increasingly frequent global crises that are leading us towards civilisational collapse. The Confucian philosophy of yin and yang is the antithesis of this way of thinking. Peace with nature, learning from it and imitating it are the guarantees for the future of our civilisation.

Following the ideas of this book, a new humanism is emerging, linked to the search for balance, sustainability, and disaster resilience: a new enlightenment that demands solidarity, humility, and respect for Mother Earth and future generations. This is, in short, our message fully aligned with the ideas of the Club of Rome. Club of Rome (von Weizsäcker and Wijkman, 2018).

References

Boulding, K. (1966) The Economics of the Coming Spaceship Earth. In: Jarrett, H., Ed., Environmental Quality in a Growing Economy, Resources for the Future/Johns Hopkins University Press, Baltimore, 3–14.

Rockström, J., et al. (2009a). Planetary boundaries: Exploring the safe operating space for humanity. *Ecology and Society, 14*(2), 32. http://www.ecologyandsociety.org/vol14/iss2/art32/.

Rockström, J., Steffen, W., Noone, K., Persson, A., Chapin, F. S., 3rd, Lambin, E. F., Lenton, T. M., Scheffer, M., Folke, C., Schellnhuber, H. J., Nykvist, B., de Wit, C. A., Hughes, T., v der Leeuw, S., Rodhe, H., Sörlin, S., Snyder, P. K., Costanza, R., Svedin, U., Falkenmark, M., Karlberg, L., Corell, R. W., Fabry, V. J., Hansen, J., Walker, B., Liverman, D., Richardson, K., Crutzen, P., & Foley, J. A. (2009b, September 24). A safe operating space for humanity. *Nature, 461*(7263), 472–475. https://doi.org/10.1038/461472a.

von Weizsäcker, E.U., & Wijkman, A. (2018). *Come on!: Capitalism, short-termism, population and the destruction of the planet.* Springer Science + Business Media LLC. ISBN-13: 978-1493974184. http://www.ub.edu/prometheus21/articulos/obsprometheus/BOULDING.pdf.

Printed in the United States
by Baker & Taylor Publisher Services